Pioneers in Arts, Humanities, Science, Engineering, Practice

Volume 21

Series editor

Hans Günter Brauch, Mosbach, Germany

More information about this series at http://www.springer.com/series/15230
http://www.afes-press-books.de/html/PAHSEP.htm
http://www.afes-press-books.de/html/PAHSEP_Rogers.htm

Paul Rogers

Paul Rogers: A Pioneer in Critical Security Analysis and Public Engagement

With a Foreword by Jenny Pearce

Springer

UNIVERSITY of BRADFORD

Paul Rogers
Department of Peace Studies
University of Bradford
Bradford, West Yorkshire, UK

Series Editor's Acknowledgement: The cover photograph as well as all other photos in this volume were taken from the personal photo collection of the author who also granted the permission for publication in this volume. A book website with additional information on Paul Rogers and his major book covers is at: http://www.afes-press-books.de/html/PAHSEP_Rogers.htm.

ISSN 2509-5579 ISSN 2509-5587 (electronic)
Pioneers in Arts, Humanities, Science, Engineering, Practice
ISBN 978-3-030-06985-8 ISBN 978-3-319-95150-8 (eBook)
https://doi.org/10.1007/978-3-319-95150-8

Copyediting: PD Dr. Hans Günter Brauch, AFES-PRESS e.V., Mosbach, Germany

This Springer imprint is published by the registered company Springer International Publishing AG part of Springer Nature
The registered company address is: Gewerbestrasse 11, 6330 Cham, Switzerland

For Amber and Felix

Foreword

Security and Peace: Building a Creative Interface

Introduction

Over five decades, Paul Rogers has explored the varied aspects of global insecurities. The selection here illustrates the range and depth of these contributions. From trade, development and food security to weapons technologies, in particular the nuclear, to the phases of Britain's defence posture; to the growing power of sub-state armed actors and the vulnerability of modern cities to asymmetric warfare. Most recently, Rogers has touched the nerve of the twenty-first century by showing the connections between inequality, the environmental limits of human activity and the rise of a new range of 'threats from the margins' for which traditional military solutions have no adequate answer.

Though evidently important in their own right, Rogers' attention to these particular aspects of global insecurity should not overshadow the real significance of his body of scholarship to more universal debates. Rogers has meticulously researched each of the tangled threads that constitute and reshape global security policy and thinking over time. Simultaneously, however, he has unravelled their logics and demonstrated how successive generations of politicians and policy makers in the West appear to have been incapable of making our world more secure. Indeed, at the time of writing, the headline is that NATO acknowledges that the world is at the greatest risk for a generation. This emerged from a Guardian interview with NATO's secretary general, Jens Stoltenberg (*Guardian*, 9/09/2017), who referred to the sheer number of converging threats. Rogers has been tracking our journey to this point for five decades. However, he has done so with a compass. My comments are about how that compass has oriented him to build what I call a creative interface between security and peace.

The Journey and the Compass

Security Seen from the Global South
Rogers' journey begins with his early foray as a young biologist into the 'developing' world, what we now call the Global South. From that starting point of the late 1960s, the compass takes shape. As a Latin Americanist myself, deeply imbued by experiences in that region which enabled me to see the world through the lens of the vulnerable, the dominated and the excluded, I appreciate how Rogers' compass was, at least partly, forged through an encounter with what he calls the 'margins' in Africa. The margins are both a physical location in relationship with a centre, and a state of mind of individuals who feel their powerlessness in varied social relationships. When you live with people in both senses of marginality, of place and social space, you realise that far from powerless, people are often extraordinarily resilient and creative. Their leverage to generate change in their favour, however, is limited. Rogers early on saw the significance of this, when he studied food security and the global international economic order of the 1970s and its failure to prevent famine, despite overall world sufficiency in grain reserves. Rogers' ability to 'see from the margins' has enabled him to understand the frustrations of poverty as unfairness rather than a natural condition of the planet. It is possible to trace this early exposure to the lens of marginality to his empathy for the way some on the margins read the policies and discourses of the 'centre', particularly the USA and its allies in the West.

Thus, it is not the indifference to poverty *per se* which lies behind the chronic failure of Western policy towards the Middle East and elsewhere in the Global South. Rather, it is the relational aspects of marginality that have been misunderstood by those enjoying the privileges of absolute domination over the future of peoples and planet. Blindness to the slow fuse of frustration is what Rogers exposes as his work moves from the marginality engendered by the 'New Economic Order' of the 1970s, to the marginality within the global economic, social, military and political order of the unipolar American world of the post-Cold War era.

Access to education in the global South, which as Rogers describes, has been an important feature of more recent history, can enhance the sense of marginality when it fails to offer a dignified future. The effects are not inevitably lethal, of course. The vast majority channel frustrations into varied forms of resilience and resistance. However, when combined with the way so many cultures foster the performance of masculinity, the effects can be deep on the psyche of young men who cannot fulfill expectations. These might range from that of providing for the family to the need for status and recognition and often includes the search for dominance and power over others. The evidence that over two thirds of homicides are committed by young men on young men suggests that we need to understand much better this variable without crude reductionism. In the recent history of the Middle East, North Africa and elsewhere, for instance, male and female frustrations have been channelled into democratising movements as in the Arab Spring. Many men and women have made desperate migratory journeys in search of the means to life. The perverse and dangerous side has been taken up by a minority, as when young men in particular have turned to Jihadist terrorism. When they do, the lethality and destruction have been devastating.

Rogers' compass has enabled him to enter into the mindset of marginality, and to recognise its complex expressions. He has used his insights to write a series of reports *as if* he was part of a fictional Al-Qaeda Strategic Planning Cell, and letters from Raqqa *as if* he was a young male ISIS militant. His compass oriented him to communicate—not to justify—the psychic impact of marginality and the unprecedented kind of security threat it has generated in certain very specific contexts.

Security from the Cold War to the Post-Cold War

Rogers traces the efforts of the West to secure its military dominance during and after the Cold War. His focus on nuclear options and their alternatives opened up to scrutiny the assumptions of NATO in the 1980s that maintaining a policy of early first use of nuclear weapons in the event of catastrophic reversals in a conventional conflict with the Soviet Union could be justified. Similarly, he explores the justifications behind the UK commitment to a nuclear option through Trident, despite eschewing the first use option. Rogers does not endear himself to the defence establishment through these contributions, although he has been frequently asked to give talks to them.

With an acute sense of temporality, Rogers reminds us that the generations under their 40s today do not live under the shadow of nuclear war. They are more influenced by the narratives of the Cold War which saw nuclear weapons as a deterrent. Rogers' journey through time, navigated with his peace compass, reminds us of the contingency around the roles of these weapons. More than ever before, we need to listen to those who argue that rather than deterrence, nuclear war was only very narrowly averted and we came so much closer than anyone dares to recognise during the Cold War. In addition, the Cold War was played out in the theatres of the Global South, such as El Salvador, where I experienced it directly in the 1980s, when the US government in the name of 'preventing communism', played a huge role in preventing a genuine social as well as political transformation in that small and vulnerable country. As the threat of nuclear proliferation grew in the post-Cold War, the short-sightedness of assuming that nuclear weapons can keep the world safe is clear. However, it is only through activists and scholars such as Rogers, who have navigated with a peace compass the varied phases of their development (alongside other chemical and biological weapons), that we can see the continuities and risks in the discourses of justification by the powerful.

The same compass is apparent as Rogers' insights in the early years after the Cold War revealed how blind were the powerful to emergent threats on their doorstep and their meanings for the future of their own and global security. Already in the 1990s, Rogers saw the devastation (mostly to infrastructure at this stage) of the Provisional IRA's crude bombs in the cities of London and Manchester. He highlighted the significance of the attack on the New York Trade Centre in 1993, which could have opened minds to the kind of violent attack within the belly of a nation, that the West might not even have a concept for yet. It was not long before 9/11 provided that with the 'war on terror'. However, as Rogers' documents so well in his book, *Irregular Warfare*, the West, in particular the USA and the UK, deluded themselves that interventions in Afghanistan and Iraq would quickly contain the new problem and return the world to US hegemony in a market-driven global economy.

Security and the Conditions to Live without Violence

Throughout these contributions, Rogers brings together the world of security policy and the real world it is trying to contain. He is showing the disjuncture between these. As the post-Cold War era enters into the crises of the twenty-first century, Rogers is urging his readers to accompany him on this journey through the complex and apparently disparate threads of security policy, and to see the looming dangers ahead to which it appears to offer no direction. Only, I argue, with Rogers' compass, can these dangers be made visible. Only if one recognises that the *conditions* to live without violence have to be prioritised, can one appreciate the dangers looming on the horizon as well as their immediate expressions. Of course, their immediate expressions are of huge concern. However, an inability to analyse and respond to them through a compass of peace means that military solutions continuously prevail, however much they reproduce more violence, consolidate inequalities and avoid the ultimate shared threat to everyone on the planet of climate change. A military-as-security response has become a kind of 'common sense'. It is this common sense that Rogers' unpacks so skilfully.

Conclusion

Rogers is a natural scientist, who has migrated to social sciences. His concerns are material realities not ephemeral ideals. His compass is thus not a series of abstract norms, but a principled realism. In the dialogue between security and peace studies, Rogers has freed himself from the constraints of disciplinary silos and what I have termed this 'common sense' of militarised security paradigms. In so doing, he has created the intellectual potentiality of what it might mean to see security in terms of a more peaceful world not one managed in the interests of a few people and places. Much of Rogers' writings are about war, but I would argue, they are actually about peace. They highlight how a security approach that focusses on war can actually fail to see the new expressions of collective violence emergent in its entrails. Rogers' peace compass has enabled him to refine peace thinker, Johan Galtung's, famous dichotomy between negative and positive peace. He shows that it is the interface between the two that matters. Security policy must be seen in constant dialogue with the task of building the conditions to live without violence. It cannot reproduce more violence. It must recognise the political impacts of inequality as well as their social ones. A peace policy for security is urgent but evidently will not emerge overnight. Without it as a compass, however, we return again and again to mutating expressions of violence in the social relations we construct.

Over the last century, contributions to peace have ranged from Einstein to Malala Yousafzai, from a male German Jewish physicist to a young female Pakistani Muslim educationalist. The circle is widening for peace, and Rogers has done a great deal to widen it further. Having had the fortune of working with Rogers in Bradford Peace Studies, I can affirm that he widens the circle in multiple ways, of which the written word readers have in front of them is only one. A commitment to communicating, in accessible ways, in varied forms and to multiple publics (including the military) with an 'alternative common sense' marks Rogers out within academia. This helped shape the Department as one where the compass of peace navigates our journeys of scholarship as well as our role as citizens.

London, UK
September 2017

Prof. Jenny Pearce[1]
London School of Economics
Caribbean and Latin America Centre

[1] Jenny Pearce is Research Professor at the Centre for Latin American and Caribbean Studies at LSE and was, from 1992 to 2016, in the Department of Peace Studies at University of Bradford.

Acknowledgements

From Potatoes to Peace

I would like to thank the staff of Springers for their work in seeing this through to completion and the several publishing houses for giving permission for the inclusion of previously published articles. More generally, this volume is due almost entirely to the commitment of Hans Guenter Brauch as series editor and it would not have been developed without his hard work throughout. I would also like to thank Oliver Ramsbotham for his key contribution to one of the publications in this collection, and Jenny Pearce for writing such a thoughtful preface which taught me a lot about what I have been trying to do over the past fifty years.

During that time, I have worked with many different groups in the UK and other countries and have learnt hugely from the interactions and sharing of knowledge and experience. Staff and students in the Peace Studies community at Bradford University are remarkable in their commitment and breadth of knowledge and it has been a hugely valued learning experience to have been part of that community over the past forty years. More personally, I would like to thank my wife, Claire, and our sons, Robert, Thomas, Edward and William for bearing with me throughout—not easy at the best of times.

When Hans Guenter approached me a couple of years ago about including this volume in the series, my initial reaction was to say 'no', since I was not at all sure that the work merited this kind of presentation and could readily be seen as a bit of a vanity project. Having thought about it and talked to friends who took a much more positive view, it did seem that there might be some value to it, for three different reasons.

One is that my own career path has been thoroughly odd, starting with a doctorate in the microbial ecology of soil-borne plant pathogens of root crops (common scab disease of potatoes) and ending up in peace studies specialising in research into responses to political violence. It is true that a number of peace researchers have had careers that have involved major changes, but mine is a bit extreme even by those standards. That might give hope to others engaged in the eclectic pursuits so common to peace research and who wonder where they might end up.

The second is that I have tried throughout the whole period to combine academic work with a commitment to engaging in public debate on the issues and that experience does suggest that it is readily possible to do so. The print and broadcast media are weird beasts, and if you are outside the mainstream, then it is not always an easy matter to get your ideas across. Mike MccGwire once told me that the trick was to be able to say outrageous things acceptably. I am still working on that but it surely makes good sense to keep trying.

Finally, what has been both really challenging and thoroughly worthwhile has been trying to make some kind of sense of a worldwide predicament that has so many issues to address and so much new thinking needing to be done. Coming to terms with a global community that is facing pervasive socio-economic marginalisation and environmental limits but with elite communities emphasising the need to maintain the *status quo* means that alternative responses are essential. I have been privileged for most of my career to work in academic environments that still enable people to spend some of their time on their own work, with little limitation as to what they study. With that comes a responsibility to use the time to good effect, and if this collection of articles shows some evidence of that, then it will have served its purpose.

Kirkburton, Yorkshire, UK
May 2018

Paul Rogers

Contents

Part IV Selected Texts on British Defence Policy

Part V Selected Texts on Global Political Violence

Part I
On Paul Rogers

Paul Rogers aged 7 with his mother Eileen and father Lawrence, Chingford, August 1950. *Source* From the author's personal photo collection

Chapter 1
Biographical Background

Paul Rogers was born in Chingford, a suburb in north east London, on 10 February 1943. His father, Lawrence, was a firefighter during the war in the London Fire Brigade and his mother, Eileen, had taught elocution before the war. He had a sister, Mary, two years older, and another sister, Lucy, was born five years later. Paul was too young to remember the war, but has clear memories of the devastation that was still visible across the East End of London and the docklands after the war. Although Chingford was an outer suburb that was away from the main target areas of the docks and industry, there was a very large anti-aircraft artillery battery close to the town, and bombers caught in the bombardment would frequently jettison bombs on the town. It was also hit by V-1 cruise missiles and V-2 ballistic missiles towards the end of the war.

After the war Paul's father developed a small wholesale ironmongery business which he ran mainly on his own, supplying shops across east London and north Kent. The family was Roman Catholic by belief and Lucy, who had severe learning difficulties, was taken to Lourdes in the hope of a cure. It wasn't to be and she died quite young. His mother had long periods of depression and both she and his father died when Paul was in his twenties.

Paul went to a local state school, McEntee Technical school in Walthamstow, and then on to Imperial College at the University of London in 1961. There he took a degree in biological sciences, studying botany, zoology and chemistry, specialising in his final year in plant pathology before going on in 1964 to get a post as a research assistant in plant pathology, also at Imperial College.

His research led to a Ph.D. which was in the microbial ecology of soil-borne plant pathogens and included farm-based field trials. In 1968, at the age of 24, he was appointed to a lectureship in plant pathology in the same department. This was a post funded as part of the UK's international development programme, and the terms of the appointment were that he could be seconded overseas for the majority of the time. He continued with that work for three years, including two years as a Senior Scientific Officer with the East African Agriculture and Forestry Research Organisation, based in Kenya for a few months and then spending most of the period in Uganda.

P. Rogers, *Paul Rogers: A Pioneer in Critical Security Analysis and Public Engagement*, Pioneers in Arts, Humanities, Science, Engineering, Practice 21,
https://doi.org/10.1007/978-3-319-95150-8_1

While a student at Imperial College Paul was active in Catholic student politics at a time of ferment in Catholic social teaching, not least in relation to liberation theology. He also got heavily involved in the early campaigning on international development with groups that went on to coalesce as the World Development Movement. This is now Global Justice Now, which is the UK's largest and most active development activism group. He was also a member of the Haslemere Group which did the first substantive activist-orientated analysis of North-South trading relations and their link to the grossly unequal distribution of wealth. The Haslemere Declaration, published in 1968, provided motivation for many campaigners, and Paul learnt much from his involvement in the group. During this time he worked with many other young development campaigners, one of whom was Claire Skellern who was also researching towards a Ph.D. at Imperial College.

His period in East Africa was hugely influential on him, further stimulating his interest in the politics of development, with a particular concern with food security that stemmed partly from his professional work in a regional crop improvement research programme. This was another considerable learning experience, not least as he travelled many tens of thousands of miles by road throughout Kenya, Uganda and Tanzania engaged in field research on crop pest and disease problems. The three governments had jointly established a programme to breed varieties of sugar cane specifically suited to local climatic and soil conditions rather than rely on varieties imported from other sugar cane growing areas of the world. The aim was to satisfy increasing local demand rather than export refined sugar but it was essential that new varieties had natural disease and pest resistance. Paul's job was to determine the most important pests and disease and then establish a programme to test the newly bred varieties, ultimately handing over to a Ugandan plant pathologist who had recently completed postgraduate training.

Paul Rogers at 25 on the Uganda-Sudan border, December 1968. *Source* From the author's personal photo collection

When he returned to the UK in 1970, he and Claire got married and they moved to live in her home village of Kirkburton in West Yorkshire. By this time his academic interests had already moved towards the politics of resource use rather than just the scientific elements and Paul took a post at one of the new and pedagogically innovative polytechnics. He worked at Huddersfield Polytechnic (now a university) for eight years, being heavily involved in a radically new degree programme that integrated the study of politics, economics, ecology, geography and human health studies into a single degree programme in Human Ecology. Over that period he edited four books, co-authored a biology textbook with Claire and wrote widely on environmental issues, especially those concerned with resource conflict. Much of this revolved around the concept of environmental limits to growth and came to the fore internationally on the occasion of the first UN Conference on the Human Environment in Stockholm in 1972.

During this time his earlier interest in North-South trading relations, which had included some published articles in the late 1960s, was further developed by the advent of "producer power" through the actions of the oil producing organisation, OPEC, in 1973–74. The surge in commodity prices led in turn to calls for a New International Economic Order through the UN Conference on Trade and Development (UNCTAD) and a Special Session of the UN General Assembly in 1974. During the same period of international economic and political ferment there developed a world food crisis which resulted in an emergency UN World Food Congress in Rome in November 1974. Paul attended the conference representing the World Development Movement. Writing for their magazine enabled him to gain a press pass which, in turn, gave him something of an inside view of the complex politics of international food policy.

Claire and Paul Rogers on holiday, Norfolk, July 1971. *Source* From the author's personal photo collection

At the end of 1978 and after eight years at Huddersfield, a senior lectureship was advertised at the Department of Peace Studies at Bradford University, then a small department that had been established just five years earlier. He was to stay there in full-time posts for the next 32 years, being awarded a professorship in 1992. From 1993 to 1999 he was head of the Department and also spent a period as Chair of the University's Social Sciences Faculty and Deputy Chair of the University's Research Committee. He was also active in the British International Studies Association and in 2002 he was elected to a two-year term as Chair of the Association.

When Paul joined the Peace Studies Department in 1979 it was just as renewed East-West tensions were coming to the fore, with successive leadership crises in the Soviet Union combining with more forceful political attitudes in key western countries, most especially after the election of President Ronald Reagan in the United States and Prime Minister Margaret Thatcher in the United Kingdom. In the very early 1980s a number of new nuclear weapons systems were being developed, including US cruise and Pershing II mobile missiles being deployed in Western Europe and Soviet SS-20 mobile missiles being deployed in Eastern Europe. The heightened tensions were exacerbated by crises in the Middle East and the Soviet military occupation of Afghanistan, with one result being the growth of mass-movement anti-nuclear organisations, especially in Western Europe.

During this period Paul worked with two other colleagues in Bradford, Malcolm Dando and Peter van den Dungen, to research and write on nuclear issues. These frequently provided a critical perspective on security trends and were rarely popular with governments. They were also subject to critical scrutiny by other academics who might take a more positive view of nuclear developments. The scrutiny proved to be a very useful aid to improving the empirical and analytical quality of the work at Bradford. This period during the 1980s enabled Paul to study and analyse wider military developments, a task further stimulated by Britain's war with Argentina over control of the Falklands/Malvinas islands. While often taking a markedly independent if not critical position on defence and security issues he received regular invitations from 1982 onwards to lecture at Britain's senior defence colleges. This is a connection that continues to this day and has also involved many briefings to the UK Ministry of Defence, the Foreign and Commonwealth Office, the Home Office, the Cabinet Office and national police, security and intelligence organisations.

The family at Kirkburton in July 2006. From left: Rob and Pauline, Will, Tom, Claire, Ed, Paul. *Source* From the author's personal photo collection

Within a year of the ending of the Cold War, the 1990–91 Iraq crisis and war broke out and Paul was heavily engaged in a critical analysis of the war and its early aftermath. From 1992 onwards, and in light of an evolution in the Northern Ireland conflict involving the Provisional IRA's (PIRA) change of strategy towards economic targeting of the City of London, he began a study of the evolution of paramilitary movements, especially in relation to their strategies and tactics. By the mid-1990s he had developed a view of trends in international security that pointed to the risk of what might be termed "revolts from the margins" in a world in which widening socio-economic divisions combined with environmental limits to growth would lead to an unstable security environment in which smaller paramilitary organisations might have an impact out of all proportion to their apparent capabilities.

In many ways this was a continuation of his earlier work linking development and environment issues, not least interdisciplinary work with Tony Vann back at Huddersfield Polytechnic in the early 1970s. His book, *Losing Control: Global Security in the 21st Century,* written in the late 1990s and published a year before the 9/11 arracks, developed this analysis further, pointed to the risk of catastrophic attacks and predicted accurately how the United States and other western states would respond. After the attacks the book sold widely and was translated into a number of languages including Chinese and Japanese.

Paul working on barn roof, 1994 and the finished barn. He started building it in 1988 towards the end of the Cold War. He thought it would take three years but it took him ten, by which time a new global disorder had emerged. *Source* From the author's personal photo collection

In the decade after 9/11, Paul published extensively on the evolving war on terror while remaining at Bradford University, the published output ranging from journal articles and books through to substantial web publishing. While he retired from his full-time post at Bradford at the age of 67 in 2010 he was asked to stay on for six more years on a part-time basis before taking up an emeritus professorship in 2016 where he retains a continuing commitment to the work of the Department, including some teaching. He combines this with an association with the London-based think tank, the Oxford Research Group, which works on non-military responses to security challenges, and also with the web journal, *Open Democracy*.

For the whole of the past half century he has combined an academic career with extensive writing and lecturing in support of peace and development activism, contributing well over a thousand articles, many of them in recent years being web-based outputs with a world-wide readership. These include over 850 weekly articles for www.opendemocracy.net and close to 150 monthly analyses for the Oxford Research Group. He continues to take up frequent invitations to speak at senior defence colleges and is an Honorary Fellow of the UK's Joint Services Command and Staff College. Over three decades he has provided extensive informal advice to political leaders and has also served as an expert witness for the defence in court cases involving peace activists taking direct action at nuclear and other installations.

Paul's academic output has included 29 books and over 150 papers and chapters, and his work has been translated into many languages including Russian, Catalan, Turkish, Farsi, French, German, Polish, Italian, Spanish, Portuguese, Dutch and Greek. Since 1974 Paul has been a regular contributor to radio and television stations and networks throughout the world, including BBC, ITV, Al-Jazeera, CNN, CBC, US National Public Radio, Radio France International, RTE Dublin, FM4 in Vienna, ABC, Monocle 24, RTV Hong Kong and Radio New Zealand. He has done thousands of interviews and still averages over 200 each year.

Paul and Claire have four sons, Robert, Thomas, Edward and William, and five grandchildren. All four sons took science or engineering degrees and are now scattered around Europe, currently working in the UK, the Netherlands and Switzerland. Claire and Paul still live in the same house in Kirkburton and have a few acres of land where they grow a range of fruit and vegetables and keep some livestock. Claire worked for many years as a biology lecturer and was an examiner for the Oxford and Cambridge schools examination board for twenty-five years. She has long been active in local politics including time as a case-worker for a Labour member of parliament. She is also a bee-keeper. Although both are now non-theists they are regular change ringers – a form of church tower-bell ringing that originated in England around four hundred years ago. Paul remains a keen amateur builder and is also an occasional hill-walker.

Paul's approach to international security might best be expressed by a summary of his most recent book, *Irregular War: the New Threats from the Margins* (London: I B Tauris, 2017):

ISIS, al Qaida, Boko Haram, Al Shabab and the Taliban are all separate manifestations of a new non-state dynamic which is now driving international conflict through asymmetric and hybrid warfare, but their real significance is much more fundamental. The problem for the future is not a clash of civilisations but revolts from the margins with ISIS, in particular, having been a proto-movement for wars in an increasingly divided and constrained world. The underlying drivers of future conflict are far more than the growth of extreme Islamist movements. They stem from a deeply flawed world economic system that is producing greater inequalities and marginalisation, combined with the onset of persistent global environmental limits, especially climate disruption.

The failed war on terror shows that the consequences of these drivers cannot be controlled by military force and we cannot close the castle gates. What is required is a fundamentally new approach to security if we are to avoid a highly unstable and violent world - an age of insurgencies which might even involve weapons of mass destruction. We need radically to change our understanding of security, a change that is possible but requires vision and commitment.

Chapter 2
Paul Rogers' Publications (1963–2017)

2.1 Books

Rogers, Paul (editor), *The Education of Human Ecologists,* Charles Knight, London, 1973.

Vann, Anthony and Paul Rogers (editors), *Human Ecology and World Development,* Plenum Press, New York and London, 1974.

Rogers, Paul (editor), *Future Resources and World Development,* Plenum Press, New York and London, 1976.

Jones, J. Owen and Paul Rogers (editors), *Human Ecology and the Development of Settlements,* Plenum Press, New York and London, 1976.

Skellern, Claire and Paul Rogers, *Basic Botany,* Macdonald and Evans, London, 1977.

Rogers, Paul, Malcolm Dando and Peter van den Dungen, *As Lambs to the Slaughter: The Facts About Nuclear War,* Arrow Books, London, 1981.

De Kaunay, Bertrand, Paul Rogers, Malcolm Dando and Peter van den Dungen, *Le Poker Nucleaire,* Syros, Paris, 1983 (French adaptation of *As Lambs to the Slaughter*).

Rogers, Paul, *Guide to Nuclear Weapons,* Bradford University, Bradford, 1983.

Dando, Malcolm and Paul Rogers, *The Death of Deterrence,* CND Publications, London, 1984.

Randle, Michael, April Carter, Dan Smith, Owen Green, Paul Rogers and others, *The Politics of Alternative Defence,* Grafton Books, London, 1987 (the second report of the UK Alternative Defence Commission, chaired by Paul Rogers).

Rogers, Paul, *Guide to Nuclear Weapons,* Berg Press, Oxford and Providence RI, 1988.

Landais-Stamp, Paul and Paul Rogers, *Rocking the Boat: New Zealand, the United States and the Nuclear-Free Zone Controversy,* Berg Press, Oxford and Providence RI, 1989.

P. Rogers, *Paul Rogers: A Pioneer in Critical Security Analysis and Public Engagement*, Pioneers in Arts, Humanities, Science, Engineering, Practice 21,
https://doi.org/10.1007/978-3-319-95150-8_2

Randle, Michael and Paul Rogers (editors), *Alternatives in European Security,* Dartmouth Publishing Company, Aldershot, 1990.

Rogers, Paul and Malcolm Dando, *The Directory of Nuclear, Biological and Chemical Arms and Disarmament 1990.* Tri-Service Press, London, 1990.

Rogers, Paul and Malcolm Dando, *A Violent Peace: Global Security after the Cold War,* Brasseys, London, Washington and New York, 1992.

Tansey, Geoffrey, Katherine Tansey and Paul Rogers (editors), *A World Divided: Militarism and Development After the Cold War,* Earthscan, London and St Martin's Press, New York, 1994.

Rogers, Paul, *Losing Control: Global Security in the 21st Century* Pluto Press, London, 2000 (second edition 2002, third edition 2010).

Brauch, Hans Guenter, P.H. Liotta, Antonioa Marquina, Paul Rogers and Mohammad El-Sayed Selim (editors), *Security and Environment in the Mediterranean: Conceptualising Security and Environmental Conflicts,* Springer, 2004.

Rogers, Paul, *A War on Terror: Afghanistan and After,* Pluto Press, London, 2004.

Rogers, Paul, *Iraq and the War on Terror,* Oxford Research Group International Security Report, Oxford, 2004.

Rogers, Paul, *Iraq and the War on Terror, 2003–04,* Oxford Research Group, Oxford, 2004.

Rogers, Paul, *Iraq and the War on Terror, 12 Months of Insurgency, 2004–05,* I.B. Tauris, London, 2005.

Rogers, Paul, *A War Too Far: Iraq, Iran and the New American Century,* Pluto Press, London, 2006.

Ritchie, Nick and Paul Rogers, *The Political Road to War with Iraq,* Routledge, London, 2006.

Rogers, Paul, *Into the Long War,* Pluto Press, London, 2007.

Rogers, Paul, *Towards Sustainable Security: Alternatives to the War on Terror,* Oxford Research Group, London, 2007.

Abbott, Chris, Paul Rogers and John Sloboda, *Beyond Terror: The Truth About the Real Threats to our Security,* Rider, London, Sydney, Auckland, Johannesburg, 2007.

Rogers, Paul, *Why We're Losing the War on Terror,* Polity Press, Cambridge and Malden, MA, 2008.

Rogers, Paul, *Global Security and the War on Terror: Elite Power and the Illusion of Control,* Routledge, Abingdon and New York, 2008.

Rogers, Paul, *Irregular War: ISIS and the Threat from the Margins,* I.B. Tauris, London, 2016.

Rogers, Paul. *Irregular War: The New Threat from the Margins,* I.B. Tauris, London, 2017 (Revised edition of *Irregular War*, 2016.)

2.2 Papers, Chapters and Major Conference Reports

Rogers, P. "Imperial College Sierra Leone Expedition 1963", *Nature,* **197**, 1256, 1963.

Rogers, P. "UNCTAD 2", *The Tablet,* **222**, 6663, 97–8, 3 February 1968.

Rogers, P. "Organic Manuring for Potato Scab Control and Its Relation to Soil Manganese", *Annals of Applied Biology,* **63**, 3, 371–8, 1969

Rogers, P. "Disease Testing in Sugar Cane", *Proceedings of the 13th Meeting of the East African Specialist Committee on Agricultural Botany,* Kampala, Uganda, 1969.

Rogers, P.F. (editor), *Proceedings of the Special Meeting on the Yellow Wilt Condition of Sugar Cane,* East African Agriculture and Forestry Research Organisation, Nairobi, Kenya, 1969.

Rogers, P.F. and S.G. Mukasa, "Sugar Cane Pathology: Crop Loss Assessment Trials", *East African Agriculture and Forestry Research Organisation Record of Research 1969*, Nairobi, Kenya, 1969.

Rogers, P. "Research on Yellow Wilt of Sugar Cane", *Sugarcane Pathologists Newsletter,* International Society of Sugarcane Technologists, 1969.

Rogers, P.F., M. Simbwa-Bunnya, E. Mayangja-Kasirye, and S.G. Mukasa, "Sugar Cane Breeding: Crop Loss Assessment Trials", *East African Agriculture and Forestry Research Organisation Record of Research 1970*, Nairobi, Kenya, 1970.

Rogers, P.F. and M. Simbwa-Bunnya, "East African Sugar Cane Disease and Pest Survey", *East African Agriculture and Forestry Research Organisation Record of Research 1970*, Nairobi, Kenya, 1970.

Rogers, P. "Current Programmes in Sugar Cane Disease Research", *Sugarcane Pathologists Newsletter,* International Society of Sugarcane Technologists, 1970.

Rogers, P.F. "Chemical Control of Common Scab of Potato Tubers", *Acta Horticulturae,* **21**, 132–136, 1971.

Rogers, P.F. "Sugar Cane Diseases of Potential Importance in East Africa", *East African Agriculture and Forestry Journal,* XXXVII, 2, 93–99, 1971.

Rogers, P.F., M.M. Simbwa-Bunnya and M.P. Early, "A Survey of Sugar Cane Diseases and Pests in East Africa", *East African Agriculture and Forestry Journal,* XXXVIII, 2, 162–169, 1972.

Rogers, P. "Human Ecology in the Commonwealth", *Occasional Paper Number XIV,* The Commonwealth Foundation, London, 1972.

Rogers, P. "The 2nd Commonwealth Conference on Development and Human Ecology", *The Environment This Month,* July 1972.

Rogers, P.F. "A guide to courses in human ecology", *The Ecologist,* **2**, 5, 30–31, 1972.

Rogers, P.F. "The Education of Human Ecologists", *Commission on Education Newsletter,* 9, 7–10, International Union for the Conservation of Nature and Natural Resources, Morges, Switzerland, 1972.

Rogers, P.F. "Report on Two Human Ecology Conferences", *International Journal of Environmental Studies,* **4**, 303–306, 1973.

Rogers, Paul. "Ecology and the Environmental Crisis", *Yorkshire Architect,* **29**, 611–612, 1973.

Rogers, Paul and Anthony Vann, "Symposium on Human Ecology and World Development", *International Journal of Environmental Studies,* **5**, 309–310, 1974.

Rogers, Paul. "Human Ecology", *Occasional Paper Number XXXV,* The Commonwealth Foundation, London, 1975.

Rogers, Paul. "Third World Studies and British Environment Education", British Association for the Advancement of Science Annual Meeting 1975, Section L Education, 1975.

Rogers, Paul. "Grain – for the rich world's animals or the poor world's people?" *Briefing Paper 1/75,* World Development Movement, London, 1975.

Rogers, Paul. "Producer Power – from the oil crisis to UNCTAD IV", *Briefing Paper 7/75,* World Development Movement, London, 1975.

Rogers, Paul, *Food in our time – but not just yet: a report on the world food congress*, World Development Movement, London, 1975.

Rogers, Paul. "Human Ecology", chapter in: Martin, George C. and Keith Wheeler (editors), *Insights into Environmental Education,* Oliver and Boyd, Edinburgh, 1975.

Rogers, Paul. "The New Economic Order – implications for development and human ecology", *Fourth Commonwealth Conference on Human Ecology and Development,* Dacca, Bangladesh, 1976.

Rogers, Paul. "Political Undercurrents in Alternative Technology", *Christian Action,* 6–7, Winter 1978.

Rogers, Paul. "Economic and Social Influences", chapter in: Carson, Sean McB. (editor), *Environmental Education, Principles and Practice*, Edward Arnold, London, 1978.

Rogers, Paul. "International Resource Conflict", *International Conference on Human Ecology,* Jaipur, India, 1979.

Rogers, Paul. "War of the Resources", *Peace News,* 6–7, 8 April, 1980.

Rogers, Paul, "Global Overkill: a guide to nuclear weapons", *Peace News,* 4 pp, 15 May 1981.

Rogers, Paul. "British Overkill: a guide to British nuclear weapons", *Peace News,* 4 pp, 1981.

Rogers, Paul. "Trends in weapons and military strategy", *Futures,* **13**, 5, 428–432, 1981.

Rogers, Paul. "Britain's Nuclear Weapons Arsenals", chapter in: Bledowska, Celina (editor), *War and Order – Researching State Stuctures,* Junction Books, London, 1983.

Rogers, Paul. "Unilateral contributions to nuclear disarmament", *ADIU Report,* **5**, 1, 15–18, 1983.

Rogers, Paul. "Introduction", chapter in: Dalyell, Tam, *Thatcher's Torpedo: The Sinking of the 'Belgrano',* Cecil Woolf, London, 1983.

Rogers, Paul. "Rapid Deployment Forces and Third World Intervention", *Development Studies Association,* Bradford, September 1984.

Rogers, Paul. *A Guide to Britain's Nuclear Weapons,* 36 pp, CND Publications, London 1985.

Rogers, Paul, "Controlling the Violent Peace – Force Projection in the 1980s", *International Peace Research Association*, Brighton, Sussex, 1986.

Rogers, Paul. "Introduction", chapter in: Dalyell, Tam, *Thatcher: Patterns of Deceit,* Cecil Woolf, London, 1986.

Rogers, Paul. "The Implications for the United Kingdom of Ballistic Missile Defence", *House of Commons – Second Report of the Defence Committee 1986–87,* 147–156, June 1987.

Dando, Malcolm and Paul Rogers, "Why SALT II Failed", *ADIU Report,* **8**, 4, 4–8, 1986.

Rogers, Paul. "Force Projection and Third World Militarization", *Militarization in the Third World,* Canadian Institute for International Peace and Security, Ottawa, January 1987.

Rogers, Paul. "Fortress Falklands", *Political Studies Association,* Aberdeen, April 1987.

Randle, Michael, April Carter, Dan Smith, Owen Greene and Paul Rogers, *Alternative Defence Commission, The Politics of Alternative Defence: A role for a Non-Nuclear Britain,* Paladin, London, 1987.

Rogers, Paul. "US Military Options in the Gulf", *Peace Studies Briefing Number 32,* University of Bradford, August 1987.

Rogers, Paul. *Over Here – The US Military Presence in Britain,* CND Publications, London, 1988.

Rogers, Paul. "Cutting the Cloth", *New Statesman,* 10–11, 9 February 1988.

Rogers, Paul. "Fewer nukes don't mean a safer world", *New Statesman,* 10–11, 8 April 1988.

Rogers, Paul. "The Political Impact of the Process of Arms Control and Disarmament", *House of Commons, Third Report of the Foreign Affairs Committee, 1987–88,* 105–109, August 1988.

Rogers, Paul. "The Nuclear Connection", *Bulletin of the Atomic Scientists,* **44**, 7, 20–23, September 1988.

Rogers, Paul and Dan Plesch, "Less means more", *New Statesman,* 8–9, 21 October 1988.

Chalmers, Malcolm, Paul Rogers and Malcolm Dando, "Trident Options", *New Socialist,* 18–21, October/November 1988.

Rogers, Paul and Malcolm Dando. "No Surrender", *New Statesman,* 16–17, 8 December, 1988.

Rogers, Paul. "New Bombs for Old", *Sanity,* 20–21, December 1988.

Rogers, Paul. "Eastern Block: solutions to the Gulf crisis", *Marxism Today,* 3, September 1990.

Rogers, Paul and Malcolm Dando, "British Defence Policy in a Changed Europe", *House of Commons, Tenth Report of the Defence Committee 1989-90,* 97–102, July 1990.

Rogers, Paul. "Defence thinking – confronting a shrinking market", *GTI Electrical and Electronic Engineering Journal,* 41–42, 1990–1991.

Rogers, Paul. "Defensive Behaviour", *Marxism Today,* 27, October 1991.

Rogers, Paul. "The Myth of the Clean War", *Covert Action Information Bulletin,* 37, 26–30, 1991.

Rogers, Paul. "British Defence Policy in a Changed Europe", chapter in: Woodhouse, Tom (editor), *Peacemaking in a Troubled World,* Berg, New York and Oxford, 1991.

Rogers, Paul. "A New Order for Which World? Ethics, Culture and Intervention After the Cold War", Paper for: *Sovereignty and the Right to Intervene,* Norwegian Institute of International Affairs, May 1992.

Rogers, Paul. "Trident – Bedrock of Britain's Defence into the 21st Century?", *The Cost of Trident – Issues for Local Authorities,* Glasgow, 10 June 1992.

Rogers, Paul. "US lays plans for victory in age of North-South War", *The Observer,* 28 June 1992.

Rogers, Paul. "The Myth of the Clean War", chapter in: Crary, Jonathan and Sanford Kwinter (editors), *Incorporations,* Zone, New York, 1992.

Rogers, Paul. "Crazy for Oil", *New Internationalist,* 10–11, October 1992.

Rogers, Paul. "A Crowded, Glowering Planet", *Dialogue and Resistance,* 2–3, Autumn 1992.

Rogers, Paul. "The Falklands War and British Defence Policy", chapter in: Danchev, Alex (editor), *International Perspectives on the Falklands Conflict: A Matter of Life and Death,* St Martin's Press, New York, 1992.

Rogers, Paul. "Militarism, Poverty and International Security", Paper for: *International Workshop on Environment and Poverty,* Global Forum on Environment and Poverty, Dhaka, Bangladesh, July 1993.

Rogers, Paul. "Le rôle de l'Europe dans le nouvel ordre mondial", *Bulletin Pax Christi,* **36**, 3, 23–24, June 1993.

Rogers, Paul. "Global Security After the Cold War", *Sheffield Papers in International Studies,* University of Sheffield, 1993.

Rogers, Paul. "Trends in US Defence Policy after the Cold War", Paper for: The British International Studies Association, University of Warwick, 1993.

Gamlen, Elizabeth and Paul Rogers, "U.S. Reflagging of Kuwaiti Tankers", chapter in: Rajaee, Farhang (editor), *The Iran-Iraq War: The Politics of Aggression,* University Press of Florida, Gainesville, FA, 1993.

Rogers, Paul. "Jungle of Snakes", *New Internationalist,* **35**, February 1994.

Rogers, Paul. "Proliferation in Perspective", *Sanity,* **8**, Spring 1994.

Rogers, Paul. "Scramble for Supremacy", *New Statesman,* 20–21, 9 September 1994.

Rogers, Paul. "A Note on the British Deployment of Nuclear Weapons in Crises – with particular reference to the Falklands and Gulf Wars and the purpose of Trident", *Lobster,* 28, 2–10, December 1994.

Rogers, Paul. "A Jungle Full of Snakes? Power, Poverty and International Security", chapter in: Tansey, Geoff, Kath Tansey and Paul Rogers (editors), *A World Divided – Militarism and Development After the Cold War,* Earthscan, London, 1994.

Rogers, Paul. "The UN Role in Arms Control", chapter in: *Challenges Facing the United Nations*, Childers, Erskine (editor), St. Martin's Press, New York, and CIIR, London, 1994.
Rogers, Paul and Simon Whitby, "Conventional Weapons of Mass Destruction – trends and control", *House of Commons – Report of the Foreign Affairs Committee 1994–1995,* 308–316, March 1995.
Rogers, Paul. "Security Means More than Defence". *Parliamentary Brief*, **3**, 5, March 1995.
Rogers, Paul. "The Current Status of the Nuclear Arms Race" *Medicine and War*, **11**, 3, 1995.
Rogers, Paul. "The Spectre of Mass Destruction Terrorism", *Parliamentary Brief*, **3**, 6, April 1995.
Rogers, Paul. "The Challenge of Peace: Militarism or the Peace Economy?", chapter in: Race, Alan and Roger Williamson (editors), *True to the Earth,* One World Publications, Oxford and Rockport, MA, 1995.
Rogers, Paul. "New US weapon destroys electrics but not people", *New Statesman,* 9 June 1995.
Rogers, Paul. "The Evolution of an Interdisciplinary Peace Studies Centre: The Bradford Experience", *Peace and Conflict Studies,* **2**, 1, pp. 44–48, June 1995.
Rogers, Paul. "The Curse of the Undead", *Sanity,* 10–11, Summer, 1995.
Rogers, Paul, "The Role of British Nuclear Weapons After the Cold War". *BASIC Research Report 95.1*, British American Security Information Council, Washington DC and London, 1995.
Rogers, Paul, "Nuclear Disarmament – Pressure For and Against", *Disarmament Diplomacy*, **1**, 2, February 1996.
Rogers, Paul. "Bomb Culture", *New Statesman,* 18–19, 16 June 1995.
Rogers, Paul. "Economic Targeting and Provisional IRA Strategy", *Studies in Political Violence,* 96.1, pp 55, July 1996.
Rogers, Paul. "What are the Threats to Britain's Security?", chapter in: Gittings, John and Ian Davis (editors), *Rethinking Defence and Foreign Policy,* Spokesman, Nottingham, 1996.
Rogers, Paul, Simon Whitby and Stephen Young, "Nuclear Futures: The Role of Nuclear Weapons in Security Policy", *BASIC Research Report 96.1,* British American Security Information Council, Washington DC and London, 1996.
Rogers, Paul. "Nuclear Weapons – Disarmament, Control or Transition?", chapter in: *Issues in Peace Research, 1995–96*, Lee-Anne Broadhead (editor), University of Bradford, 1996.
Rogers, Paul. "Sub-Strategic Trident, a Slow-Burning Fuse", *London Defence Papers Number 34*, Centre for Defence Studies, London, 1996.
Rogers, Paul. "Reviewing Britain's security", *International Affairs,* **73**, 4, pp. 655–669 (1997).
Rogers, Paul. "A story of peace studies", *The Friend,* 7–8, 28 November 1997.
Whitby, Simon and Paul Rogers, "Anti-crop Biological Warfare – Implications of Iraqi and US Programs", *Defense Analysis,* **13**, 3, 303–318, 1997.
Rogers, Paul. "Characteristics of Natural Outbreaks of Crop Diseases", chapter in: Dando, M. et al. (editors), *Scientific and Technical Means of Distinguishing*

Between Natural and Other Outbreaks of Diseases, 49–61, Kluwer Academic Publishers, 1998.

Rogers, Paul. "Conflict Prevented or Postponed? Reflections on the Nearly War in Iraq", *Disarmament Diplomacy,* 24, 2–7 March 1998.

Rogers, Paul. "Near War, Real Fear", *Covert Action Quarterly,* 11–16, Spring 1998.

Rogers, Paul. "Responding to Western Intervention – Conventional and Unconventional Options", *Defense Analysis,* **14**, 1, 41–54, 1998.

Rogers, Paul. "Security Consequences of the Osiraq Raid", *Contemporary Security Policy,* **19**, 2, 115–117, 1998.

Rogers, Paul, "Strategic Directions, Strategic Choices", in *The Outcome of the Strategic Defence Review,* RUSI Whitehall Papers No. 44, Royal United Services Institute, London, 1998.

Rogers, Paul. "New Ground, Old Assumptions: Analytical Limitations in the SDR", *Disarmament Diplomacy,* 28, 4–5, 28 July 1999.

Rogers, Paul. "Kosovo Crisis – Lessons to Learn", *The World Today,* **55**, 8/9, 4–6, August/September 1999.

Rogers, Paul. "Learning from the Cold War Nuclear Confrontation", chapter in Alan P. Dobson (editor), *Deconstructing and Reconstructing the Cold War,* Ashgate, 1999.

Rogers Paul, Simon Whitby and Malcolm Dando, "Biological Warfare Against Crops", *Scientific American,* **280**, 6, pp. 70–75, 1999.

Rogers, Paul. "Analysis and Reporting of Plant Disease Epidemics", *NATO Advanced Workshop: BTWC Security Implications of Human, Animal and Plant Disease Epidemiology,* Bucharest, Romania, June 1999.

Rogers Paul and Oliver Ramsbotham, "Then and Now: Peace Research – Past and Future", *Political Studies,* **47**, 4, 1999.

Rogers, Paul. "Some Implications of Kosovo", *New Political Economy,* **5**, No. 1, 1999.

Rogers, Paul. "New Challenges to Western Security", *World Reference Systems 2000,* pp. 67–69, RUSI, London, 2000.

Rogers, Paul. "Understanding the Adversary: The Evolution of Civil Conflict", in Malcom Dando (editor), *Non-Lethal Weapons: Technological and Operational Prospects,* Jane's Information Group, London, 2000.

Rogers, Paul. "The Use of Economic Targeting by the Provisional IRA, 1992–1997, *Civil Wars,* **3**, No. 4, pp. 1–30, 2000.

Rogers, Paul. "International Security in the Early Twenty-First Century", *ISIS Briefing Paper No 22,* International Security Information Service Europe, February 2000.

Rogers, Paul. "Towards an Ideal Weapon? Military and Political Implications of the Airborne and Space-Based Lasers", *Defense Analysis,* **17**, 1, 73–88, 2001.

Rogers, Paul. "Review Essay: New Analyses of Global Challenges", *International Studies Review,* **3**, 1, 142–145.

Rogers, Paul. "Conflict and Peace in the Early 21st Century", *RUSI Journal,* **146**, 2, 9–14.

Rogers, Paul. "A Violent Peace", *The Ploughshares Monitor,* **22**, 3, 10–12, 2001.

Rogers, Paul. "European and United States Attitudes to International Security – Implications for East Asia", paper for *The International Conference on Korean Reconciliation and Reunification for Global Peace,* Seoul, South Korea, August 2001.

Rogers, Paul. "Political Violence and Asymmetric Warfare", *US-European Forum on World Order and Global Issues,* Centre for European Reform, Berlin, October 2001.

Rogers, Paul and Scilla Ellworthy, "The United States, Europe and the Majority World after 11 September", *Briefing Paper,* Oxford Research Group, October, 2001.

Rogers, Paul. "Towards an Ideal Weapon? Military and Political Implications of the Airborne and Spacer-based Lasers", *Defense Analysis,* **17**, 1, 73–88, 2001.

Rogers, Paul. "Some Implications of National Missile Defence", *Defence Studies,* **1**, 1, 167–171, Spring, 2001.

Rogers, Paul. "Resource Issues", chapter in: Salmon, Trevor C. (editor), *Issues in International Relations,* Routledge, London and New York, 2002.

Rogers, Paul and Scilla Ellworthy, "A Never-Ending War? Consequences of 11 September", *Briefing Paper,* Oxford Research Group, March 2002.

Rogers, Paul. "The United Nations and the Promotion of Peace", *Medicine, Conflict and Survival,* **18**, 110–119, 2002.

Rogers, Paul. "Limitations on Joint Warfare: the Impact of the Proliferation of Conventional Weapons Technology", chapter in *The Changing Face of Military Power: Joint Warfare in an Expeditionary Era,* Andrew Dorman, Mike Smith and Matthew Uttley (editors), Palgrave Macmillan, pp. 45–72.

Rogers, Paul. "Right for America, Right for the World", *The World Today,* **58**, 2, 13–15, 2002.

Rogers, Paul. "The War on Terror – One Year On", *RUSI Journal,* 28–32, October 2002.

Rogers, Paul "Political Violence and Global Order", chapter in *Worlds in Collision: Terror and the Future of Global Order,* Booth, Ken and Tim Dunne (eds), Palgrave Macmillan, pp. 215–225, 2002.

Rogers, Paul. "Arms control and disarmament in the context of developing countries", chapter in: Desai, Vandana and Robert B. Potter (editors), *The Companion to Development Studies,* Arnold, London, 2002 (2nd edition 2008).

Elworthy, Scilla and Paul Rogers, "The 'War on Terrorism': 12-Month Audit and Future Strategy Options", *Briefing Paper,* Oxford Research Group, September 2002.

Rogers, Paul. "Iraq: Consequences of a War", *Briefing Paper,* Oxford Research Group, October 2002.

Rogers, Paul. "War and its Aftermath – Destructive Force", *The World Today,* **59**, 5, 4–5, May 2003.

Rogers, Paul. "The 'War on Terror': Winning or Losing", *Briefing Paper,* Oxford Research Group, September 2003.

Rogers, Paul. "President Bush's War on Terror", *Disarmament Diplomacy,* 74, 1–4, December 2003.

Rogers, Paul. "'Losing Control' – War and the Modern World", *International Relations,* **17**, 1, 97–108, 2003.

Rogers, Paul. "The US Military Posture: 'A Uniquely Benign Imperialism'?", chapter in: Panitch, Leo and Colin Leys (editors), *The Socialist Register 2004,* 12–31, Merlin Press, London, 2003.

Rogers, Paul. "International Security After 9/11 – trends in international conflict 2005–2030", *Analysing Conflict and its Resolution,* Institute for Mathematics and its Applications, Oxford, 2004.

Rogers, Paul. *The Second Bush Administration and US Policy Towards Iran,* Institute for Political and International Studies, Tehran, March 2005.

Rogers, Paul. "Endless War, The Global War on Terror and the New Bush Administration", *Briefing Paper,* Oxford Research Group, March 2005.

Rogers, Paul. "The Global War on Terror: Current Status and Trends", *in The Global War on Terror: Its Genesis, Implications, Remedial and Counter Measures,* Institute of Regional Studies, Islamabad, August 2005.

Mearsheimer, John, J. Paul Rogers, Richard Little, Christopher Hill, Chris Brown and Ken Booth, "Roundtable: The Battle Rages On", *International Relations,* **19**, 3, 337–360, 2005.

Rogers, Paul. "Happy Christmas War Isn't Over", *The Parliamentary Monitor,* 135, 70–71, January/February 2006.

Rogers, Paul. "The global war on terror: four years on", *Medicine, Conflict and Survival,* **22**, 1, 4–12, January–March 2006.

Rogers, Paul. "The Global Agenda", *Contemporary Threats to International Security,* 59–61, Project Ploughshares, Waterloo, Ontario, 2006.

Rogers, Paul. "Iran: Consequences of a War", *Briefing Paper,* Oxford Research Group, February 2006.

Abbott, Chris, Paul Rogers and John Sloboda, "Global Responses to Global Threats, Sustainable Security for the 21st Century", *Briefing Paper,* Oxford Research Group, October 2006.

Rogers, Paul. "A Necessary War", *Political Studies Review,* **5**, 25–31, 2007.

Rogers, Paul. "Reconsidering the War on Terror", *RUSI Journal,* **152**, 4, 42–35, August 2007.

Rogers, Paul. "Peace Studies", chapter in: Collins, Alan (editor), *Contemporary Security Studies,* Oxford University Press, Oxford, 2007 (2nd edition 2010, 3rd edition 2013, 4th edition 2016).

Rogers, Paul. "Terrorism", chapter in: Williams, Paul D. *Security Studies, an Introduction,* Routledge, London and New York, 2008 (2nd edition, 2013, 3rd edition 2018).

Rogers, Paul. "Iraq, Afghanistan – Has the US Lost?", *The Socialist Review,* 10–13, March 2008.

Rogers, Paul. "Global Terrorism", chapter in: Cox, Michael and Doug Stokes, *US Foreign Policy,* Oxford University Press, Oxford, 2008 (2nd edition, 2012).

Rogers, Paul. "'More Muddled than Confused': A Response to David Martin Jones and M.L.R. Smith, 'Confused Britannia: Global Uncertainty and Homeland Insecurity'", *Studies in Conflict and Terrorism,* **31**, 6, June 2008.

Rogers, Paul. "Talking Warheads", *New Internationalist,* 14–15, June 2008.

Rogers, Paul. "2008, The Year in Review", chapter in: Lewis, D.S. and Wendy Slater (editors), *The 2009 Annual Register: World Events,* ProQuest, Bethesda MD and Cambridge UK, 2009.

Rogers, Paul. "Global Security after the War on Terror", *Briefing Paper,* Oxford Research Group, November 2009.

Rogers, Paul. "Climate Change and Security", *Clinical Medicine,* **9**, 2, 108–109, April 2009.

Rogers, Paul. "Energy Security and Iran's Role in International Relations", *Inaugural Farabi Conference – Iran and the International System,* Durham, June 2009.

Rogers, Paul. "International Threats, National Security", *RUSI Monitor*, **8**, 1, 6–7, August 2009.

Rogers, Paul. "2009, The Year in Review", chapter in: Lewis, D.S. and Wendy Slater (editors), *The 2010 Annual Register: World Events,* ProQuest, Bethesda MD and Cambridge UK, 2010.

Rogers, Paul. "Reviewing Britain's Security", *Briefing Paper,* Oxford Research Group, May 2010.

Rogers, Paul. "Incidents in Fallujah", chapter in: Bechler, Rosemary, (editor), *The Convention on Modern Liberty,* Open Democracy, London, 2010.

Rogers, Paul. "2010, The Year in Review", chapter in: Lewis, D.S. and Wendy Slater (editors), *The 2011 Annual Register: World Events,* ProQuest, Bethesda MD and Cambridge UK, 2011.

Zala, Benjamin and Paul Rogers, "The 'Other' Global Security Challenges: Socioeconomic and Environmental Realities after the War on Terror", *RUSI Journal,* **156**, 4, 26–33, 2011.

Rogers, Paul. "The 'War on Terror' and International Security", *Irish Studies in International Affairs,* **22**, 1–9, 2011.

Rogers, Paul and Benjamin Zala, *Written Evidence from the Oxford Research Group for the Report on the Strategic Defence and Security Review and the National Security Strategy,* House of Commons Defence Select Committee, August 2011.

Rogers, Paul. "Nonviolent Protest", *Resurgence,* 267, 26–27, July/August 2011.

Rogers, Paul. "The Strategic Defence and Security Review – missing the point?". *SGR Newsletter,* 39, 7, Winter 2011.

Rogers, Paul. "2011, The Year in Review", chapter in: Lewis, D.S. and Wendy Slater (editors), *The 2012 Annual Register: World Events,* ProQuest, Bethesda MD and Cambridge UK, 2012.

Rogers, Paul. "Chances for Peace in the Second Decade – What is Going Wrong and What We Must Do", *ORG Pamphlet*, Oxford Research Group, December 2012.

Rogers, Paul. *The World in 2012 – Trends in International Security,* CIDOR Yearbook, CIDOR, Barcelona, (published in Catalan) 2012.

Rogers, Paul. "Wars of terror – Learning the Lessons of Failure", chapter in: Jackson, Richard and Samuel Justin Sinclair (editors), *Contemporary Debates on Terrorism,* Routledge, London and New York, 2012 (2nd edition, 2017).

Rogers, Paul. "2012, The Year in Review", chapter in: Lewis, D.S. and Wendy Slater (editors), *The 2013 Annual Register: World Events,* ProQuest, Bethesda MD and Cambridge UK, 2013.

Rogers, Paul. "Security by Remote Control, can it work?" *RUSI Journal,* **158**, 3, 14–20, June/July 2013.

Rogers, Paul. *A Note on the Wider Islamist Context of Recent Developments,* House of Commons Select Committee on Foreign Affairs – North and West Africa Inquiry, May 2013.

Rogers, Paul. "Lost Cause: Consequences And Implications of the War on Terror", *Critical Studies on Terrorism,* 1–13, 2013.

Rogers, Paul. "2013, The Year in Review", chapter in: Lewis, D.S. and Wendy Slater (editors), *The 2014 Annual Register: World Events,* ProQuest, Bethesda MD and Cambridge UK, 2014.

Rogers, Paul. *A Note on the Origins, Evolution and Prospects for the Islamic State Movement,* Evidence to the House of Commons Select Committee on Defence, 2014.

Rogers, Paul. "A Fresh Start for War Weary Iraq", *The Diplomatist,* **2**, 10, 32–33, October 2014.

Rogers, Paul. "A Century on the Edge: from Cold War to hot world, 1945–2045", *International Affairs,* **90**, 1, 2014.

Rogers, Paul. "Chances for Peace in the Second Decade – What is Going Wrong and What We Must Do", chapter in: Zelter, Angie (editor), *World in Chains,* Luath Press, Edinburgh, 2014.

Rogers, Paul. "Afghanistan, Pakistan and the War on Terror", chapter in: Houen, Alex, *States of War since 9/11 – Terrorism, Sovereignty and the War on Terror,* Routledge, London and New York, 2014.

Rogers, Paul. "2014, The Year in Review", chapter in: Lewis, D.S. and Wendy Slater (editors), *The 2015 Annual Register: World Events,* ProQuest, Bethesda MD and Cambridge UK, 2015.

Rogers, Paul. "Peace Research and Conservation Conflicts", chapter in: Redpath, Stephen M. et al., *Conflicts in Conservation – Navigating Towards Solutions,* Cambridge University Press, Cambridge, 2015.

Rogers, Paul. "Climate Conflict", *Pan-European Networks, Science and Technology,* 17, December 2015.

Rogers, Paul. "2015, The Year in Review", chapter in: Lewis, D.S. and Wendy Slater (editors), *The 2016 Annual Register: World Events,* ProQuest, Bethesda MD and Cambridge UK, 2016.

Rogers, Paul. "A Critical Perspective on the Global War on Terror", chapter in: Jackson, Richard (editor), *Routledge Handbook of Critical Terrorism Studies*, Routledge, London and New York, 2016.

Rogers, Paul. "2016, The Year in Review", chapter in: Lewis, D.S. and Wendy Slater (editors), *The 2017 Annual Register: World Events,* ProQuest, Bethesda MD and Cambridge UK, 2017.

The family in Cornwall, March 2017. Adults from left: Jess, Will, Pauline, Tanya, Ed, Rob, Claire, Paul, Ksenia, Tom. Children from left: Zoe, Ben, Felix, Charlotte, Amber. *Source* Photograph by Ksenia Chmutina from the author's personal photo collection

Chapter 3
The Publications in Context: An Introduction

3.1 Development and Environment

Although much of Paul's writing over the past 30 years has sought to integrate analyses of security, economic and environmental issues, before that his main interest was in elements of trade, development and food security, stemming partly from his involvement in development activism in the early and mid-1960s followed by working in East Africa around the end of the decade. Two short article stem from that period.

3.1.1 UNCTAD 2 (1968)

The Tablet is a British magazine that was first published in 1840 and reports from a Catholic perspective, having a reputation for covering a wide range of theological and social issues.[1] In the early 1960s, the increasing membership of the UN General Assembly drawn from recently independent ex-colonies was one of the major factors in establishing a new agency devoted primarily to the links between trade and development. The main focus was on the manner in which colonial-era North/South trading patterns disadvantaged such countries. With a small permanent secretariat based in Geneva, the UN Conference on Trade and Development (UNCTAD) held its first negotiating session in that city in 1964, with the second planned for early 1968 in New Delhi.

By that time, the UN Development Decade of the 1960s was drawing to a close with little prospect of reaching its various targets. From the position of a socially aware journal such as *The Tablet*, this was a concern, and Paul wrote an editorial for

[1]This text was published as: "UNCTAD 2", *The Tablet*, 3 February 1968 and is republished with permission.

P. Rogers, *Paul Rogers: A Pioneer in Critical Security Analysis and Public Engagement*, Pioneers in Arts, Humanities, Science, Engineering, Practice 21,
https://doi.org/10.1007/978-3-319-95150-8_3

the paper that addressed the risks of failure. It cited the recently agreed Charter of Algiers from a large group of what were then known as less developed countries. This was rooted in the idea that if world trading patterns were intrinsically biased in favour of the more powerful industrialised states then that should be countered by mechanisms such as commodity agreements, tariff preferences and compensatory finance systems. This was radically different from what later came to be known as the neoliberal approach. The editorial concluded by suggesting that the UK might have a particular role to play under its then Labour Government, given that it had been significant in preventing the failure of UNCTAD 1.

In the event, very little was achieved at UNCTAD 2 in New Delhi but only five years later North/South trading relations were thrown into disarray with the emergence of OPEC as a major player and the quadrupling of world oil prices between October 1973 and May 1974, coupled with a general rise in commodity prices. This altered the terms of trade in favour of the primary commodity exporting nations which meant that many of the UNCTAD ideas were taken up in the hastily convened Sixth Special Session of the UN General Assembly in early 1974. This resulted in the Declaration on a New International Economic Order, the core of which was an Integrated Commodities Programme that would do much to regularise North/South trade to the greater advantage of poorer commodity producers.

Within a year, though, recession had set in after the oil price rises, commodity prices fell, the powerful trading states of the Global North no longer had the motive to negotiate and little was achieved. By the end of the 1970s, the neoliberal model was being progressively adopted, to be reinforced by the collapse of the Soviet-era centrally planned model at the end of the 1980s.

3.1.2 Prospects for Food (1975)

Following his attendance at the UN World Food Congress in November 1974, Paul wrote an article for another religious journal, *The Month*, on the manner in which a world food crisis had emerged in the early 1970s and the response that had followed.[2] It detailed the transition from optimism over world food supplies at the end of the 1960s with the apparent success of the "green revolution" and the new high-yielding cereal varieties through to the threat of major famines by 1972/3. Exploring the complex of long-term and short-term factors involved in the change, the article pointed to the synergistic nature of the interactions as a core factor behind the deterioration in food prospects.

It went on to point to the short-term failure of the Congress to respond to the immediate needs but, written some 12 months after the Congress, it could also point to the impact of the publicity as encouraging some states, including newly-wealthy

[2]This text was first published as: "Prospects for Food", *The Month*, October 1975 and is republished with permission.

oil producers, to provide emergency aid, this doing much to alleviate the immediate risk of famine. At the time of writing, the New International Economic Order still showed promise, but the risk of failure was also apparent. Perhaps the most significant aspect of the whole food crisis period of the early 1970s was that there was never an occasion when overall world grain reserves were insufficient for world-wide needs. The problem was of shortfalls in food production in some areas with local and regional populations too impoverished to buy food to make up the difference. The fundamental problem, and the potential cause of famine at the time, was poverty not overall shortages of food.

3.2 Cold War and After

For more than a decade from 1979 Paul worked mainly on nuclear issues and more general aspects of East-West security relations. During this period he became very interested in a number of developments in military technology, primarily related to nuclear issues, and began regular lecturing at a number of UK defence colleges. He was also involved in the second phase of the Alternative Defence Commission, acting as chair of the commission as it continued its work on proposing alternative defence strategies for existing nuclear powers such as the UK.

Two articles published around the end of that decade and one relating to the same era but published later, are examples of this interest. The first explores the implications of the Strategic Defence Initiative, the second stems from the Alternative Defence Commission and benefits from the considerable knowledge of Michael Randle and April Carter, and the third looks back at the nuclear dimension of the Cold War and tries to draw lessons. Given the resurgence of concern over East-West relations in the late 2010s and the failure to get even remotely close to a nuclear-free world, there is continuing salience to all three aspects of that era.

3.2.1 The Transformation of War: The Real Implications of the SDI (1989)

Shortly after his election in 1980, President Ronald Reagan had reached the conclusion that the rate of expansion of Soviet strategic nuclear forces was a fundamental threat to the security of the United States but that there was considerable potential for the development of anti-missile defences, even against individual warhead travelling at 7,000 mph or more.[3] In 1983 he announced the development

[3]This text was first published as: "The Transformation of War: The Real Implications of the SDI", chapter in *Military Technology, Armaments Dynamics and Disarmament*, edited by Hans Günter Brauch, 1989 and is republished with permission.

of a Strategic Defence Initiative (SDI), almost immediately transformed in common parlance into the "star wars" programme. In the years that followed, many technologies were investigated, including third generation nuclear weapons where the immense energy bursts might be able to be aimed in particular directions. The idea behind this was that relatively small nuclear devices that had such capabilities might be launched into positions well above the earth's surface and detonated to direct their power to destroy incoming nuclear warheads.

A second major strand was the programme to develop powerful long-range lasers that could 'lase' ballistic missiles soon after launch and destroy them before they even left the territory of the enemy. To defenders the great advantage of this was that if the destruction of the missile lead to radioactive contamination from damaged warheads, then it would be the enemy's homeland that would be hit. Furthermore, if an intercontinental ballistic missile with multiple warheads could be destroyed early in its flight path, as many as ten warheads might be destroyed at once.

There was, though, another implication to this development which related to what might be described as a theory of weapons. Put simply, the purpose of a weapon is to deliver energy from a source to a target in order to disrupt or even destroy that target. Such a function might vary from the swing of an axe, the flight of a spear or arrow, the discharge of a gun or the use of a missile. In all cases the basic purpose is the same. When looked at in this way, an ideal weapon would be one that had unlimited range, pinpoint accuracy, travelled at the speed of light and was impossible to counter. A powerful space-based laser system could come close to this ideal in terms of speed, accuracy, range and invulnerability. As such, it might have a transforming impact on warfare, to the benefit of any state able to develop it.

This article examines just that potential, utilising some of the material on US laser programmes that had, by the late 1980s, come into the public domain. It does so in the context of the history of key weapons technologies in the closing years of the Cold War but there is a later element, covered in a further paper published ten years later which chartered the development of the airborne laser (ABL). ("Towards an Ideal Weapon? Military and Political Implications of the Airborne and Space-based Lasers", *Defense and Security Analysis*, **17**, 1, pp. 73–88, 2001.) This was substantially funded through the 1990s in the wake of the 1991 Gulf War when the United States armed forces found that the threat from Iraqi Scud missiles was far more serious than anticipated. The aim would be to destroy such missiles at launch using a powerful chemical iodine-oxygen laser mounted on a heavily modified Boeing 747. In the event, the programme was cancelled largely on grounds of cost but the technologies that were developed during this period have since been used in many other projects, the aim still having an element of seeking to get close to an "ideal weapon."

3.2.2 *Alternative Military Options in Europe (1990)*

During the final decade of the Cold War, and especially in the early 1980s, there was considerable interest in the idea of alternative defence postures for countries such as Britain that continued to depend on nuclear forces.[4] One of the sets of experiences relevant to this was related to the defence postures of three European powers, Sweden, Switzerland and Yugoslavia. The first two were neutral states that had taken steps to develop nuclear weapons but had decided not to do so, relying instead on strong conventional forces backed up by very large reserves in order to make occupation by an enemy state too costly. While the third, Yugoslavia, had not gone down the nuclear path it saw itself sandwiched between East and West and was intent on maintaining a non-aligned status partly by evolving a defence posture not unlike that of Sweden and Switzerland.

At the height of the 1980s NATO still maintained a policy of early first use of nuclear weapons if faced with potentially disastrous reversals in a conventional conflict with the Soviet Union. Such a policy was a particular spur to peace researchers and others to investigate alternative postures that emphasised defence with the aim of making a conventional assault too costly, and the book of which this chapter forms part, was a contribution to that debate.

3.2.3 *Learning from the Cold War (1999)*

This was developed from a chapter in *Deconstructing the Cold War*, edited by Alan P. Dobson (1999) and reflects on the nuclear history of the Cold War era. Its focus is partly on the nature of arms races, the potential destructiveness of nuclear war and the dangers of crises escalating into war. It also pays particular attention to nuclear weapons accidents. In all, it seeks to provide an overview of the era which included several serious crises and accidents, all at a time when the potential for nuclear first use was clear.

Reflecting on this nearly two decades later, one is struck by the relevance of a study of this period to current circumstances. While the ludicrously excessive arsenals of the Cold War era are now much diminished, there is no prospect of serious moves to general nuclear disarmament, and all the five major nuclear powers are modernising their weapons. India and Pakistan are expanding their nuclear forces and North Korea is trying to develop a small arsenal, Israel is assumed to be maintaining powerful forces and only in the case of Iran has a move towards nuclear weapons been halted, at least for now.

Moreover, from 2015 onwards there was a notable rise in East West tensions, especially between the NATO powers and Russia, and the election of

[4]This text was published as: "Learning from the Cold War". Chapter in *Losing Control: Global Security in the 21st Century*, (2000) and is republished with permission.

Donald J. Trump as US President in November 2016 ushered in a further period of uncertainty. One of the issues of concern here is that the potential for nuclear destruction, and the dangers and uncertainties of nuclear confrontation are foreign to anyone not in their mid-teens by 1990. Thus, people not yet in their mid-forties have little appreciation of the nuclear issue. It is perhaps one of the most neglected areas in international relations and even, to an extent, in peace studies.

3.3 The United Kingdom's Defence Posture

One focus of Paul's work since the late 1970s has been the UK defence posture, and this has been covered in a number of monographs, papers and book chapters. He worked extensively on the Falklands/Malvinas War in the early and mid-1980s and one of the elements of that work which attracted substantial interest was the question of whether the UK deployed tactical and strategic nuclear weapons within range of Argentina. This led to his interest in nuclear targeting and the question of whether Britain's Trident strategic nuclear missile force included a proportion of missiles with low-yield 'sub-strategic' warheads. This was explored in a monograph published by the London-based Centre for Defence Studies at King's College in 1996. Subsequent to this work Paul also wrote on a more general analysis of UK defence posture, especially the issue which came to the fore by the early 2000s – the intention to replace the fleet of Trident missile submarines while also constructing two very large aircraft carriers. His argument was that this would largely determine Britain's defence posture for several decades in a manner that would greatly limit its versatility.

By the early 1990s he had become increasingly interested in paramilitary movements, partly prompted by the manner in which the Provisional IRA (PIRA) in Northern Ireland began a campaign of economic targeting in London intended not to cause high civilian casualties but to damage the status of London as it competed with Frankfurt to be the financial hub of Western Europe. This interest developed during the decade into a more general analysis of the manner in which relatively small sub-state actors could have an impact on powerful states and even international organisations. It was an analysis which ultimately linked with his long-term interest in wealth disparities and the potential for environmental limits to growth and was to lead on to the writing of "Losing Control". The work on PIRA included a number of short articles and one substantive paper published in the journal *Civil Wars* in 2000: "Political Violence and Economic Targeting – Aspects of Provisional IRA Strategy, 1992–97", **3,** No. 4, pp. 1–30.

3.3.1 *Sub-strategic Trident: A Slow Burning Fuse (1996)*

In line with NATO nuclear policy, the UK did not embrace the idea of no-first-use of nuclear weapons and during the Cold War it maintained a range of tactical nuclear weapons.[5] These included a number of US-owned weapons such as the warheads for the Lance artillery missiles, shells for M109 and M110 guns and anti-submarine nuclear depth bombs for delivery by Nimrod maritime reconnaissance aircraft. All of these were held under a dual control system where authorisation for use would be by the agreement of both governments.

Meanwhile, the UK's strategic nuclear force of Polaris missiles on nuclear-powered submarines was replaced by Trident missiles on a new class of submarines in the 1990s and, in addition, the UK government had developed and deployed its own tactical nuclear bomb. This was the WE-177 bombs with a yield reported to be around 10 kilotons, broadly similar in destructive power to the Hiroshima and Nagasaki bombs. It was deployed on strike aircraft over many years, including Tornado, Jaguar and Buccaneer aircraft. A depth bomb variant could be carried by Sea King and Lynx helicopters deployed on aircraft carriers, destroyers and frigates.

All of the UK-deployed US weapons were withdrawn after the end of the Cold War in the early 1990s and the UK government phased out the WE-177 over the same period, leaving just the Trident system, but an option for tactical or sub-strategic use was retained through the availability of a low-yield version of the standard Trident warhead. The usual version was reported to have a destructive power of around 100 kilotons with the low-yield version likely to be similar to that of the WE-177. The maintenance of this option, in the context of eschewing a no-first-use posture was the subject of the analysis of this paper, which sought to place it in the context of a long history of the UK being capable of fighting a nuclear war, even against a non-nuclear opponent. The paper attracted considerable interest at the time and the Ministry of Defence was reported to be unhappy with its publication, perhaps not surprising given the UK government's long-standing preference for minimal discussion on the details of nuclear posture, especially as it relates to targeting.

3.3.2 *Big Boats and Bigger Skimmers: Determining Britain's Role in the Long War (2006)*

Returning to the theme of Britain's defence posture a decade later, this paper widened the discussion to explore the manner in which Britain's role in

[5]This text was published as: Sub-Strategic Trident: A Slow Burning Fuse, *London Defence Papers Number 34*, (London: Brasseys for the Centre for Defence Studies, King's Collee, London, 1996) and is republished with permission.

international security risked being severely limited by two major procurement decisions – the construction of two *Queen Elizabeth*-class aircraft carriers and four *Dreadnought*-class ballistic missile submarines.[6] The aircraft carriers would be, by some measure, the largest warships ever deployed by the Royal Navy, they would embark the new F-35 US-built strike aircraft and a number of helicopter types, and each would require destroyer and/or frigate escorts, a nuclear-powered attack submarine and logistic support ships if assembled into a carrier strike force. The argument in this paper was that almost every other part of the Royal Navy, not least its amphibious capabilities, could be squeezed out by this concentration of resources.

More generally, to sustain a ballistic missile submarine on station and ensure continuous at-sea deterrence requires considerable resources including four missile submarines, with several attack submarines and surface ships to provide "deterrence support". The argument of the paper was that this would result in what might essentially be a "two-ship navy", limiting other roles. Given that Britain has long seen itself as a maritime power, this seemed obtuse. Furthermore, if the global security environment after 9/11 suggested a very different pattern of insecurity, more important roles for Britain might relate to contributions to UN and other peace-keeping endeavours. In essence the two decisions meant that Britain was intent on maintaining its historic vision as a traditional power, but with restricted capabilities, rather than being willing to embrace responding to a new world disorder.

3.4 Global Political Violence

"Losing Control: Global Security in the 21st Century" was written mainly in 1998 and 1999 and published in 2000, and analysed broad trends in international security. Its primary arguments were, firstly, that the main determinants of conflict would relate to two potential drivers of insecurity – the widening socio-economic divide and climate disruption – and the prevalence of the control paradigm that was about elite communities maintaining the status quo, if need be by the use of military force. Related to this was the second argument that it was such an approach that would tend to incite "revolts from the margins" and that urban industrial states were vulnerable to what might best be called irregular or hybrid war. It pointed specifically to the risk of catastrophic attacks and predicted accurately how the United States would respond.

Immediately after the 9/11 attacks in 2001, Paul wrote a short paper arguing against going to war with al-Qaida and its Taliban hosts in Afghanistan. This was a radically different approach to the mainstream view at the time but has stood the test

[6] This text was published as: "Big boats and bigger skimmers: determining Britain's role in the long war", *International Affairs*, 82, 4 (2006), 651–665 and is republished with permission.

of time well. He later contributed a chapter, "Political Violence and Global Order" to *Worlds in Collision* edited by Ken Booth and Tim Dunne, developing the arguments in "Losing Control" and applying them in the new context. The conclusion was that while there was a possibility that 9/11 would prompt some new thinking on security, it was more likely that "The prognosis, at least for the present, has to be one of endless war, with a substantially increased risk of political violence, both by elites and against elites".

Even by the time that *Worlds in Collision* was published, in late 2002, there was a real prospect of a war between a US-led military coalition and the Saddam Hussein regime in Iraq. In October of that year, some five months before the outbreak of the war, Paul published a briefing with the UK think tank, Oxford Research Group, *Iraq: Consequences of a War*, pointing to probable serious consequences of a war. Once again this ran counter to most opinion within the academic security community but once again much of what it argued has proved to be highly relevant.

During this period he also developed two series of web-published articles tracking the evolution of global insecurity. These were weekly articles for Open Democracy starting in September 2001 and more lengthy monthly analyses for Oxford Research Group from May 2004. One aim of a number of these articles was to explore the world views of the extreme movements in conflict with western states. They included more than twenty reports from a fictional consultancy, the South Waziristan Institute for Strategic Hermeneutics (the SWISH Reports), mostly written for the (also fictional) Al-Qaida Strategic Planning Cell, although some were written for the US State Department and the UK government. These stretched over a ten-year period and attracted quite a wide following but, more recently, another approach was adopted, the "Letters from Raqqa" published between 2014 and 2016.

3.4.1 Responding to the World Trade Centre and Pentagon Attacks (2001)

This note was written immediately after the 9/11 attacks and privately published within six days. It was distributed to a number of policy makers and non-government organisations and a version was subsequently published on the *Open Democracy* website. The analysis was then developed jointly with the founder of Oxford Research Group, Dr. Scilla Elworthy into a substantial paper published by ORG in October 2001 (*The United States, Europe and the Majority World after 11 September*). The note argued that the 9/11 attacks should be seen as a provocation by the al-Qaida movement designed to incite a major military response from the United States and other western states. Rather than respond in this manner the note argued that the attacks should be treated as appalling examples of transnational criminality the response to which should be to take an international legal route. It was an analysis that was radically different from the norm and was almost entirely ignored at the time, but after seventeen years of war it is a view that gets a much more measured response.

3.4.2 *Political Violence and Global Order (2002)*

The paper summarised the main determinants of international security examined in *Losing Control*, not least the evolving capacity of sub-state groups to take action against centres of elite power, and then applied the analysis to the impact of the 9/11 attacks.[7] It discussed the particular political climate in the United States that followed the election of President George W Bush in November 2000, especially the influence of the neoconservative outlook on foreign and security policy. Prior to the 9/11 attacks the Project on the New American Century had been influential in promoting the view that the end of the Cold War had provided the United States with an unparalleled opportunity to provide global leadership. This would ensure that the 21st century, the "American Century", would usher in a peaceful democratic era rooted in the neoconservative vision of a truly free market and guided by the power and authority of the United States.

In such an environment 9/11 represented a devastating threat to a potentially peaceful world order overseen by the United States, and the paper argued that the inevitable result would be the vigorous pursuit of military action to counter that threat – otherwise the New American Century was at risk almost before it had begun. The paper went on to contend that this failed to understand the extent of the divisions that had developed worldwide, not least the level of anti-American feeling. As a result it concluded that 9/11 would be the start of a protracted war with no easy end in sight, a prediction sadly supported by subsequent events.

3.4.3 *Iraq: Consequences of a War (2002)*

In the months preceding the outbreak on the Iraq War in March 2003 public opposition to the war grew rapidly in many western countries, culminating in very large public demonstrations in February.[8] Most security analysts tended to accept the argument that Iraq was a major threat and that regime termination was the best option. There were exceptions one of them being this briefing from Oxford Research Group published the previous autumn. It pointed to three particular

[7]This text was published as: "Political Violence and Global Order", Chapter in: Ken Booth and Tim Dunne, *Worlds in Collision: Terror and the Future of Global Order*, (2002) Palgrave/St. Martin's Press and is republished with permission.

[8]This text was published as: "Iraq: Consequences of a War", *Briefing Paper*, Oxford Research Group, October 2002 and is republished with permission.

consequences that might ensue if military action was taken, high civilian casualties, the risk of an escalation to weapons of mass destruction and post-war regional instability. While wrong on the risk of WMD use it was correct in anticipating high civilian casualties and it also concluded, correctly, that it "should be expected to lead to an enhanced environment for paramilitary organisations to promote their policies and actions, both in the region and against the United States and its allies."

3.4.4 *Letters from Raqqa (2017)*

The "Letters from Raqqa" were supposedly from a young educated Iraqi to a friend in Baghdad.[9] From October 2014 to December 2016 ten letters tracked the progress of the conflict from a convinced believer employed by the ISIS leadership to analyse western political attitudes. At the start of the period, ISIS was controlling substantial territory in Iraq and Syria and could argue that it was a new caliphate with a population of around six million people, the primary emphasis being in further developing that caliphate. As such it had little interest in facilitating attacks on the "far enemy" in North America and Western Europe, but within two years it had come under intense pressure in a concentrated air war led by the United States.

By early 2016 it was clear that the ISIS leadership was now promoting attacks on western states, most likely for three reasons – revenge for the loss of life in the air war, to demonstrate the continuing capability of the movement and, most important, to incite anti-Muslim bigotry. The letters attracted a wide following although the very idea of trying to look at the world through ISIS eyes led to some strong criticism. They were published at the time they were 'written' but later grouped together as a single document by Open Democracy and reprinted by Oxford Research Group with an introduction in January 2017.

3.5 Putting It Together

Peace studies really developed as a field of study after the end of the Second World War. Much of the work in the 1950s and 1960s was in direct response to the tensions and dangers of the Cold War nuclear confrontation, but the concern with wider issues of peace and conflict meant that there was also a long-standing concern with other aspects of violence, conflict resolution and post-conflict peace-building. One result of this was that peace research was less thrown off course by the sudden end of the Cold War compared with international security studies as a whole.

---

[9]This text was published as: "Understanding your enemy: Donald Trump and ISIS", *International Security Briefing*, London: Oxford Research Group, January 2017 and is republished with permission.

Reflecting on this, both in terms of the Cold War period, the first decade after and possible futures was a paper co-written with Oliver Rambotham, who succeeded Paul as Head of Department at Bradford in 1999.

More recently, Paul also reviewed past and present trends and future possibilities for a more stable and peaceful world in an essay looking over the century from 1945 to 2045, putting forward the view that this would be the key century for the human community to comes to terms with both with its capacity for self-destruction and also for global ecological disaster as environmental limits to growth were exceeded. While examining hugely dangerous trends and the risks attached it also struck a positive note.

3.5.1 Then and Now: Peace Research – Past and Future (1999)

The paper argues that a number of characteristics of peace research make it distinctive and particularly suited to analysing the complex causes and consequences of violence.[10] It is a pursuit which seeks positive change towards more just and peaceful environments, it is interdisciplinary, drawing principally on the social sciences but also the humanities and natural sciences, and it maintains a focus on nonviolent principles of political and social change. It is multi-level, with an increasing focus on avoiding a "top-down" approach in favour of "peace-building from below", it intentionally seeks to be both global and cross-cultural and in doing so seeks to combine objective analysis and normative commitment with a close relationship between theory and practice, each informing the other.

3.5.2 A Century on the Edge: From Cold War to Hot World, 1945–2045 (2014)

1945 is taken is the start of this key century because of the first test and use in war of nuclear weapons, ushering in the nuclear age and a period of great danger.[11] Written in 2013–14 before an increase in east-west tensions and especially the election of President Donald J. Trump, it singled out the particular dangers of the development of weapons of mass destruction but also the manner in which there had been some scaling back of wildly excessive destructive potential after the end

[10]This text was published by Paul Rogers and Oliver Ramsbotham, "Then and Now: Peace Research – Past and Future", *Political Studies* (1999), *XLVII, 740–754* and is republished with permission.

[11]This text was published "A century on the edge: from Cold War to hot world, 1945–2045", *International Affairs*, 90: 1 (2014) 93–106 and is republished with permission.

of the Cold War. It could be argued that a global nuclear catastrophe had probably been avoided more by luck than by judgement but did point to possible developments in biotechnology and nanotechnology that could usher in new dangers.

While tolerably optimistic in this respect the article then addressed the issue of the accelerating anthropogenic impact on global ecosystem homeostasis. While this did not mean that a sudden catastrophe was probable it pointed to the very considerable dangers of not responding to the unequivocal impacts of environmental limitations, arguing that one problem was that curbing the effects, especially of climate disruption, required unpopular policy changes to be made well in advance of the maximum destructive impacts. What made the challenge worse was the interaction of global environmental limits with the deepening socio-economic divisions that had developed since the embracing of the neo-liberal economic approach in the early 1980s. The paper concluded:

> If progress is made towards a markedly less weaponized, more sustainable and more equitable world during the course of this century, then those alive towards its end may well reach the conclusion that the necessary transformation in behaviour and attitude came to the fore not just in the last third of the period discussed here, 1945-2045, but perhaps even more specifically in the late 2010s and early 2020s.

Such is the pace of change that if the essay had been written in 2018 rather than 2013–14, it might have been somewhat less optimistic.

Part II
Selected Texts on Development and Environment

Chapter 4
UNCTAD 2 (1968)

The recent public opinion poll conducted by the *Sunday Times* showed that most people considered that overseas aid was the item of government expenditure that could be most readily cut.[1] This result suggests that current efforts by the various voluntary aid organisations to educate people to an awareness of their responsibility to the less developed countries has had an effect only on a minority thinking in terms of charity rather than justice. It is a situation made more depressing by the fact that the next two months are likely to prove crucial for the future development of these nations.

The beginning of February sees the start of UNCTAD 2 in New Delhi, a meeting which will last several weeks and may well see a confrontation between poor and rich nations. The United Nations Conference on Trade and Development (UNCTAD) met for the first time in Geneva four years ago. It arose from the desire of the less developed countries for there to be an effective world forum in which could be examined the means of closing the widening gap between rich and poor nations, by the reorganisation of world trade and changes in the aid programmes of the wealthy nations.

UNCTAD 1 was not a success. Virtually the only good to come out of it was its self-perpetuation as an international organisation required to meet, every three years. Such a result may seem only to add one more to the list of top-heavy UN bureaucracies but in the case of UNCTAD this may be unjustified for two reasons. First, the problem of development is so critical that member nations are unlikely to allow vague motions and platitudes to take the place of hard bargaining. Secondly, UNCTAD already provides a valuable alternative approach to the problems of the world economy to that of the General Agreement on Tariffs and Trade (GATT).

The important role that GATT has played in world trade since the war has been based on the idea that the general reduction of tariffs and similar barriers to trade is

[1]This text was first published as "UNCTAD 2", *The Tablet*, pp. 97–98, 3 February 1968, ISSN: 0039-8837. The permission to republish this text was granted on 30 January 2018 by Philippa Lee on behalf of *The Tablet*.

P. Rogers, *Paul Rogers: A Pioneer in Critical Security Analysis and Public Engagement*, Pioneers in Arts, Humanities, Science, Engineering, Practice 21,
https://doi.org/10.1007/978-3-319-95150-8_4

to the mutual advantage of all trading countries. The UNCTAD viewpoint suggests that this GATT consent does not hold when countries of widely differing economic structure and development are concerned. According to UNCTAD free play of market forces acts against the interests of developing countries and a more controlled pattern of world trade will be necessary for them to develop at even the same rate as the wealthy countries.

The crux of the matter is that any realistic approach by the wealthy countries to the problem of world poverty will inevitably mean a lowering of their rate of increase of wealth, leading to a world in which wealth is more equally distributed. There must therefore be a basic will on the part of the wealthy countries to commit themselves to global development. At present, this will is lacking. Indeed, the position of the less developed countries is getting progressively worse both from the point of view of trade and aid.

The share that the less developed countries have of world trade has decreased from 27 per cent in 1953 to 19.9 per cent in 1965. There have been no real improvements in trade advantages to these countries, and in the case of some primary products such as cocoa, wide fluctuations in world market prices have played havoc with the economies of particular countries. The Cocoa Conference has three times broken down because of pressures from the American trade. Manufactured goods from developing countries everywhere come up against prohibitively high tariff walls, and all the Western countries restrict imports of "low-cost" cotton goods.

At the same time, the flow of aid from wealthy to poor countries has not increased. Excluding the Eastern bloc, this runs at about six thousand million dollars per year and has remained at this level for the past seven years. This means that the actual proportion of Gross National Product contributed by donor countries has decreased considerably, since, in all donor countries, GNP has been rising steadily. Actual cuts in aid have also occurred, examples being the £20 million cut in UK aid last year, devaluation, and the recent one hundred million dollar cut in the budget of the American Foreign Aid Agency.

These facts give little consolation to the delegates attending UNCTAD 2, and the likelihood of bitter clashes at New Delhi is not lessened by the fact that the less developed nations are not operating in isolation from each other. They met in Algiers last autumn and formulated proposals for UNCTAD 2 designed to narrow the poverty gap. The proposals have become known as the "Charter of Algiers" and the most important points are:

1. That commodity agreements be set up to stabilise markets and prices of primary products such as cocoa and sugar.
2. That a system of tariff preferences be established which would give free access to the markets of the developed countries to manufactures and semi-manufactures of developing countries.
3. That developed countries should reach a target by 1970 of 1 per cent of their Gross National Product in *net* financial aid.

These proposals are not in any way radical yet at the present time the chances of their acceptance by the developed nations are remote.

There is one way in which this situation might be remedied. Although it may not be possible for some Western countries to accept the Algiers Charter in the near future, it may be apparent to some that, in the long term, such proposals must be put into effect. If one or two of the wealthy nations were to commit themselves to this end then the whole tone of the conference might be changed and a positive mood result. Britain played a key part in averting the impending failure of the first UNCTAD conference four years ago. She is in a position to undertake such a role again. It would involve a real commitment on the part of the Government but could be the beginning of a new function for this country in world affairs. The President of the Board of Trade, Mr. Anthony Crosland, will be stating Britain's policy at UNCTAD in his opening speech next Tuesday in New Delhi and this will give us an idea of the role which the Government intends Britain to play.

Chapter 5
Food Security (1975)

The world food crisis in 1973 and 1974 gave considerable cause for concern about the chances of major famines affecting many countries of the third world.[1] A year after the crisis reached its peak, the world food situation remains dangerous but there is some evidence to suggest that the worst aspects of the crisis have been overcome, at least on a temporary basis. There is, however, a danger in this development in that we may become complacent to the longer term problems and, once again, be caught unawares.

In the late 1960s a degree of optimism developed among agronomists and others concerned with world food problems. The development of a number of high-yielding varieties of major food crops seemed likely to ensure a substantial increase in food production, especially in heavily populated countries such as India, Indonesia and Pakistan. This 'Green Revolution' included rice varieties produced by the International Rice Research Institute in the Philippines and varieties of wheat, maize and various beans and peas produced at other plant breeding stations in many tropical countries.

The effects of the introduction of these new varieties were certainly seen in world food production figures for two or three years, but by 1973 the scene had deteriorated sharply. This eventually led to the convening of the UN World Food Congress in Rome in November 1974, a congress which had the twin tasks of meeting an immediate crisis of food shortages and also of preparing plans to ensure adequate food supplies in the years to come.

At the time of the congress, the immediate crisis had two elements.[2] Approximately 30 less developed countries required additional grain supplies totalling about 10 million tons by June 1975. A senior UN official at the congress suggested (privately) that unless these deficits could be met, then as many as 40

[1]This text was first published as "Prospects for Food", *The Month*, pp. 276–279, October 1975. The journal does not exist any longer and no copyright holder could be located so the copyright has reverted to the author.
[2]See J.F.X. Harriott, "Facts on food", *The Month*. December 1974, pp. 800–803.

P. Rogers, *Paul Rogers: A Pioneer in Critical Security Analysis and Public Engagement*, Pioneers in Arts, Humanities, Science, Engineering, Practice 21,
https://doi.org/10.1007/978-3-319-95150-8_5

million people could starve to death, an international famine ten times as bad as the 1943 Bengal famine. The second element of the crisis was a shortage of fertilisers affecting the same group of countries and amounting to a deficit of around 1½ million tons.

How did this crisis develop, given the optimistic outlook of a few years earlier? The answer, with hindsight, is that the optimism was misplaced and that a number of unexpected factors interacted in an already hazardous situation to effect a crisis. The existing situation was one of massive poverty throughout the rural areas of the third world that produce most of the world's food, to talk of the United States and Canada as being the world's granary is misleading in that it hides the reality that two-thirds of the world's food is produced by farmers in third world countries.

The continuing causes of world poverty are beyond the scope of the present article but suffice it to say that the underlying cause of the world food crisis was the deep poverty affecting the hundreds of millions of peasant farmers who support most of the world's people.

Added to this were six further factors which all contributed to the crisis.

(1) A world fertiliser shortage which developed in the early 1970s and resulted in greatly increased fertiliser prices and also actual shortages in many third world countries.
(2) Poor weather conditions in many parts of the world in 1972, 1973 and 1974. Some climatologists have suggested that this was part of a long term shift in the distribution of rainfall throughout the world. If this turns out to be the case then some areas will no longer be capable of supporting even marginal agriculture.
(3) Pressures of population growth, still above 3 per cent per annum in many countries. As well as the additional demands this makes on food requirements, rapidly expanding populations tend to have a large proportion below the age of twelve and unable to contribute significantly to food production.
(4) Increasing food production in developed industrialised countries. Throughout the 1960s, the tendency in the richer countries of the world was towards increased food consumption especially meat. In ecological terms, meat is a relatively inefficient food in that it can take up to 10 lb weight of vegetable protein fed to an animal to produce 1 lb weight of meat. To put this in perspective, about 42 per cent of world grain production is used as animal feed, mainly in rich countries. Increases in meat consumption in these countries put marked pressures on world food supplies.
(5) The oil price rises of 1973 and 1974. These had a catastrophic effect on the economies of many third world countries. As well as the direct effects, the oil price rises stimulated inflation especially for agriculturally important commodities such as pesticides and artificial fertilisers, the latter already affected by shortages.
(6) Finally, the partial failure of the Green Revolution. Many of the apparent miracle varieties of crops needed large amounts of fertilisers, adequate irrigation water and effective provision for pest and disease control to optimise yields. Such inputs were beyond the scope of many third world farmers and the

yields were therefore not attained. Plant breeding strategies now emphasise the production of crop varieties which will out-yield older varieties under existing agricultural conditions, but it will be several years before these strategies bear fruit.

5.1 The Congress and Its Aftermath

The 12-day long world food congress had to meet immediate food shortages by ensuring that potential donors would meet the food grain and fertiliser deficits of affected third world countries. It also had to take steps to prevent future crises. Concerning such long term aims the congress sought to establish a new authoritative world food council backed up by a massive new world agricultural development fund amounting to five thousand million dollars worth of investment every year for the next ten years. A world food bank was planned which would hold 10 million tons of grain in reserve each year to meet short term shortages until the new investment programme could take effect. Finally, the much more general problem of inequalities between nations was to be tackled by a whole range of new trade and aid schemes.

The world food council was set up during the congress, but the 'teeth' of the council, the new fund, remained virtually empty by the end of the congress. The discussions on trade ended up in watered down resolutions and the plan for a food bank, though set up in theory, was unlikely to be under effective international control. The use of food aid as a political lever remained possible.

Regarding the immediate crisis, the results of the congress were even more disappointing. Only about 10 per cent of the grain and fertiliser deficits were covered by commitments made during the conference. In the case of grain, the donors were Canada and to a lesser extent Australia and Sweden, and the providers of fertilisers were Norway, Holland, Israel and Britain. But the grain committed was less than a million tons and only about 160,000 tons of fertiliser were forthcoming. Thus by the end of the congress the future for world food supplies looked bleak and many people were predicting a catastrophic year to come.

At the time of writing (mid-August) the predicted famine has not developed. Localised famines persist but, as a result of some unexpected yet welcome developments, famine on a world scale has, so far, been avoided.

A number of grain commitments made shortly after the end of the food congress resulted in the 'grain gap' standing at about 7.5 million tons at the start of 1975. By April it had fallen to less than 2.5 million tons yet less than one third of the grain had been made available to the countries in need at concessional rates. In other words, the great majority of the grain had been purchased as part of normal transactions on the grain markets.

In view of the worsening financial situation of the countries requiring the grain, it is almost certain that the great majority of the finance needed to buy the grain came from an external source and there is strong circumstantial evidence to suggest that a large part of it came from the Organisation of Petroleum Exporting Countries (OPEC) member states. A feature of international affairs that has received little attention in countries such as Britain is that the OPEC countries are rapidly becoming major aid donors. By the end of 1974, aid commitments from OPEC countries had risen to a point where they were almost equal to those of all the developed industrialised countries put together.

Of course, much of the OPEC 'aid' went to politically friendly states and relatively little of it would have benefited the poorest of the people in recipient countries. But this is also true of foreign 'aid' from the 'old rich' countries. Even so, India is believed to have received a thousand million dollars-worth of OPEC money and Pakistan and Bangladesh have also benefited substantially. The evidence certainly suggests that money from OPEC states went a long way towards filling the grain gap.

It is, however, dangerous to assume that, because the famine was avoided early in 1975. It has therefore been postponed indefinitely. A major fertiliser shortage still exists, and the FAO estimates anticipate further major grain shortfalls in the months ahead.

If the short term prospects for world food supplies look rather bleak then the long term prospects must give even greater cause for concern. At the present time, about 500 million people, almost all in third world countries, are undernourished and suffer from deficiency diseases of one sort or another. While it is true that there is still enough food to go round on a world scale, overconsumption and wastage in rich countries helps to ensure that distribution is inadequate and a whole complex of factors ensures that the countries with the food deficits cannot expand food production to meet demands.

Short of catastrophic famine, global epidemics or thermonuclear war it is highly unlikely that the world's population will stabilise at less than twice the present figure of 4000 million people. Most demographers would consider that the best prospects for stabilisation lie at a figure between 10,000 and 15,000 million around the middle of the twenty-first century. Within the next twenty five years the population of the world is likely to grow' to between 6500 million and 7000 million and it should be stressed that such an estimate takes into account recent falling birth rates in some industrialised countries as well as China and some other third world countries.

Taking these estimates along with current food shortages means that we have to nearly double food production in twenty five years. To put this task in perspective we have to remember that man first learnt to farm about 12,000 years ago. Thus, 12,000 years of improvements in agricultural productivity have to be compressed into 25 years.

There are, of course, substantial differences. There exist today a large number of agricultural developments which have not yet been put into practice on a worldwide scale. In technical terms, some people would argue that it is possible to achieve the necessary food production increases, provided that the massive financial resources which are required are actually available and also provided that there are the social and economic incentives to enable peasant farmers to utilise improved methods.

In technical terms, perhaps the greatest hope lies with four areas of development, all concerned with food crops. These are:

- the use of improved varieties,
- provision of high quality seed supplies, and
- control of pests and diseases, and adequate provision of water supplies.

The first two often go together, seed supplies ensuring that new varieties are distributed. Pests and diseases together often account for 30 per cent losses in crops before harvest with considerable further losses in storage. Finally, availability of water is all too often the limiting factor for good crop husbandry in the tropics.

Considerable scope also exists for improving animal husbandry especially in the mixed cropping of wild and domesticated animals in semi-arid pastures. Inland and marine fisheries also have potential for expansion, although in the latter case the potential may not be quite as great as is often supposed. Finally there are synthetic foods or foods of novel origin such as algae which have undeniable potential albeit frequently limited by available technology.

Mention of the technical possibilities can easily lead to false optimism and the extent of the problem to be faced must be emphasised. Above all we have to appreciate that a really revolutionary improvement in world food production is unlikely to be forthcoming without quite radical change at two different levels. The first is at the international level where a fundamental redistribution of wealth will be a pre-requisite to the funding of the agricultural development required, as well as to the eliminating of poverty as a whole.

Equally important is the essential requirement for social justice within countries. All the agricultural developments in the world will result in little progress if exploitative systems persist. To expect a peasant farmer to welcome a new development from the standpoint that he will, in turn, be exploited with greater efficiency is nothing if not naive. The slogan 'no green revolution without a red revolution' may be equally misplaced but deserves some thought.

If we return to the matter of international redistribution of wealth, then the lesson of the past two decades is that the developed industrialised countries will pay little more than lip-service to the problems of world development. In any case we are long since past the stage where we could think in the simple terms of the rich giving charity to the poor. The UNCTAD sessions as well as the more recent food congress have confirmed that, and in any case the matter was somewhat overtaken by the events which started on 17 October 1973.

On that day, the Organisation of Petroleum Exporting Countries increased the price of crude oil by 70 per cent, the first in a series of major increases. OPEC's activities ushered in a new era of producer power in which the raw materials produced in the third world are no longer necessarily cheap resources for the use of the rich. It may be too early to assess the long term trends in this phenomenon of producer power, but we can say with some certainty that things will never be the same again.

5.2 A World Turned Upside Down

Producer power is likely to have a fundamental effect on the relationships between rich and poor countries. It is closely linked with the food issue as the provision of food supplies might well become an example of producer power in the reverse direction.

The spectacular success of the OPEC moves, followed by the much less publicised but highly significant increases in the price of phosphates from North Africa, are the only two major examples of successful producer power over the past two years. But the moves towards producer power by third world countries and groups of countries extend far beyond these two resources. They include copper, iron ore, bauxite (an aluminium ore), rubber, mercury, coffee, sugar and even bananas. But more than this they extend to a series of major inter-national meetings, the results of which add up to the probable emergence of a new power bloc in the world. From the Fourth Summit Meeting of Non-Aligned Countries in Algiers in September 1973 through to the Sixth Special Session of the UN General Assembly the following April and especially to the conference in Dakar of third world resource producers earlier this year, it has become clear that a potentially powerful link between the financial resources of OPEC countries and the human and physical resources of many third world countries is likely to develop in the years ahead.

The much discussed New International Economic Order (NIEO), originally formulated at the Sixth Special Session, may indeed be in need of much refinement, but essentially it seeks a redistribution of world wealth through the medium of higher prices for third world commodities. Whether, of course, such a redistribution will ever take place must remain highly questionable. The attitudes of the rich countries of the world remain in doubt. So far the indications are that they seem unlikely to make any real concession to third world demands. We may have to wait until the next session of the UN Conference on Trade and Development (UNCTAD 4) in Nairobi next May to see whether cooperation or confrontation becomes the order of the day.

The strength on the side of the would-be exercisers of producer power is that third world countries produce significant amounts of a great many of the resources required by the rich industrialised nations. Their great weakness is that the surpluses of the one *essential* resource, food, are held by rich countries. These rich countries also have an immense technological, military and even financial superiority, but food is an astonishingly powerful political lever.

Western Europe, Australia and especially North America control a very large proportion of the world's surplus grain production, a producer power second only to oil in importance.

Of course this is a simplification. Producer power is not a single movement involving all third world countries, and many of the countries with large populations have limited capacities for exercising such power. But producer power is gathering steam and there is no doubt at all that it is receiving considerable attention within the governments of the industrialised countries.

In summary, then, in the remaining quarter of this century we are likely to sec major world food shortages. We are also likely to witness the emergence of producer power. The latter could induce fruitful co-operation between rich and poor nations in the fight to end food shortages and, indeed, overcome world poverty. The alternative is one of confrontation between rich and poor, with food supplies constituting a weapon for use by the rich in maintaining their privileged position.

Part III
Selected Texts on the Cold War and After

Chapter 6
The Transformation of War: The Real Implications of the SDI (1989)

6.1 Context

President Reagan's 'Star Wars' speech of' 23 March 1983 resulted in an intensification of US research and development into strategic defence, not the initiation of an entirely new programme.[1] The idea of constructing a defence against offensive nuclear missiles is almost as old as the missiles themselves. The United States and the Soviet Union undertook intensive research and development programmes into ballistic missile defence systems in the 1960s[2] and were also prepared to develop and deploy such systems as well as related weapons intended for use against satellites.

The difference was that in 1983, President Reagan had come to believe the Soviet offensive missile threat was awesome but was capable of being countered by strategic defences which would be effective in providing general protection to the population. Since the 1983 speech and the development of the Strategic Defence Initiative, an intense debate has developed as to whether the initiative will prove to be feasible. This debate may, however, obscure a more fundamental consideration whether the development of strategic defences forms part of a wider evolution of military technology which may have implications for the conduct of international conflict as fundamental as the development of nuclear weapons over forty years ago.

---

[1]This text was first published as: "The Transformation of War: The Real Implications of the SDI", chapter in *Military Technology, Armaments Dynamics and Disarmament*, edited by Hans Günter Brauch, published by Macmillan in 1989, which is now part of Springer Nature. This text is republished with permission of the new publisher and the editor.
[2]A useful general account of the development of ballistic missile defences, including anti-satellite weapons is Curtis Peebles, *Battle for Space* (London: Blandford, 1983).

P. Rogers, *Paul Rogers: A Pioneer in Critical Security Analysis and Public Engagement*, Pioneers in Arts, Humanities, Science, Engineering, Practice 21,
https://doi.org/10.1007/978-3-319-95150-8_6

Both the United States and the Soviet Union are involved in intensive research and development efforts directed towards strategic defence, although the US orientation is probably more towards space-based systems than that of the Soviet Union. As at present envisaged, the initiative will translate into multi-layered defence systems including boost, post-boost, mid-course and terminal phases. Systems to be deployed within a decade may rely heavily on kinetic kill systems, primarily rocket-powered. In the longer term a variety of technologies is under development which tend towards one common feature the transmission of large bursts of energy over long distances with great accuracy and speed. They are primarily but not solely space-based and the energy is likely to derive from chemical stores or controlled nuclear detonations.

Whether the various systems can combine to produce a viable strategic defence is at the centre of the current debate, although the financial commitments also loom large. The strongest supporters of SDI persist with the view that it will develop into a system able to protect an entire state or group of states, although most acknowledge that there will be a process of evolution from partial towards complete protection.[3] More measured support concentrates on the development of limited defences for military systems such as missile fields, this being seen as an aid to nuclear deterrence by making a counterforce nuclear strike less tenable.

Critics of SDI contend that comprehensive defences are virtually unattainable because of the wide range of technological innovations necessary and because of the many counter-measures which may exist. These include measures designed to increase the survivability of offensive ballistic missiles such as rapid-burn boosters, reflective or ablative coatings to missiles and manoeuvrable re-entry vehicles.[4] Alternatively, ballistic missiles might be entirely circumvented by an increased dependence on air-breathing low-altitude cruise missiles.

Whatever the merits of the arguments, the debate is normally conducted in the context of defences against ballistic missiles. In view of the recent acceleration in funding of research and development this may be considered reasonable, but this chapter argues that there is a more fundamental long-term context which should be examined that SDI, taken with developments in offensive nuclear weapons, is part of a process of producing intensely powerful weapons with low levels of collateral damage when employed against specific targets. Put simply, it will be argued that SDI forms part of a process of developing high-powered directed energy systems which are militarily usable and could transform warfare. To place this in context it is necessary to review some major features in the development of weapons.

[3]See Paul B. Stares, *Space Weapons and US Strategy* (London and Sydney: Croom Helm 1985) for a good general account of SDI and the range of arguments contributing to the debate.

[4]Discussed in Rip Bulkeley and Graham Spinardi, *Space Weapons: Deterrence or Delusion?* (Cambridge: Polity Press, 1986).

6.2 The Evolution of Weapons

A weapon is essentially a device designed to direct energy towards the destruction or incapacitation of' a target, whether that target is a physical object or a human being. This is broadly true in relation to the simplest or most complex weapon.

At the simplest level, a fist used to strike against another person involves the transmission of energy for the purpose of overcoming the person attacked. A spear thrown at an enemy in battle is a means of delivering energy over a longer distance from the body of the attacker to that of the enemy to maim or kill that person. With a bow and arrow, the energy of the drawn bow can be transmitted more accurately and over greater distances than the spear.

Spears, arrows, swords, pikes and clubs were all variations on this theme of delivery of energy, and defences could involve armour, walls and other forms of protection.[5]

A major development came with the discovery of means of using chemical energy, initially in large cumbersome guns which could propel a solid metal ball and, later, the development of small, human-portable devices such as muskets. In Europe by the sixteenth century, these weapons had developed to the point where they were having a radical effect on land and sea warfare. Large siege guns could be manufactured which could destroy massive walls to the extent that castles as the main sources of protected power became increasingly obsolete.

In the nineteenth century, a further major development involved the production of the shell, mortar bomb and grenade, all devices in which the projectile could itself contain an explosive charge which could he detonated by a fusing system after it had been delivered. The gun, mortar or howitzer became simply the means of delivering the projectile. In the early twentieth century, bombs were developed which could be delivered to a target by aeroplanes but the destruction was wrought by the release of the chemical energy of the bomb's charge.

With the exception of chemical and biological weapons, virtually all weapons follow this pattern – they direct energy to a target. The Second World War was a period of intense development, not so much in the weapons themselves as in their delivery systems, and the evolution of mass production techniques which allowed massive numbers of weapons and delivery systems to be produced.

In the past forty years there has been a veritable transformation in the development of conventional weapons, taking many forms. The 'traditional' high explosive bomb has been developed into the area impact munition, in which a single weapon contains a number of bomblets which can be dispersed and detonated over a large target. The BL755 cluster bomb is an example. One weapon of around 250 kg contains 147 bomblets, each detonating to release up to 2000 high velocity fragments, the whole munition being capable of dispersal over an area of about 0.8 ha. Fuel-air explosives have been developed which are as great an

[5]One of the best general accounts of the history of warfare is the recent overview by Gwynne Dyer, *War* (London: Guild Publishing, 1986).

advance over napalm as napalm was over Greek Fire. In these weapons, small tanklets of oxygen are incorporated in a highly inflammable gel which can thereby be detonated to produce intense but dispersed heat of great destructiveness.

Precision-guided munitions can deliver shaped-charge explosives effectively against armour, and Close Inboard Weapons Systems (CIWS) are widely in service which can fire non-explosive projectiles at up to 50 rounds *per second.* So precise are some PGMs that they are close to being able to intercept and destroy high velocity ballistic missiles. In 1986 both the United States and the Soviet Union successfully tested smart weapons against short-range ballistic missiles.

6.3 The Nuclear Transformation

Even with the many weapons developments of the mid and late twentieth century, nuclear weapons represented a transformation in the ability to transmit energy, in that they could release energy several orders of magnitude greater than was previously possible with small systems. In the 1940s, individuals within the scientific community rightly warned of the dangers of this transformation.

In the military context, however, nuclear weapons were seen as an extension of conventional weapons. The development of strategic conventional bombing against Germany and Japan resulted in the use of 1000- and 2000-bomber raids capable of delivering up to 10 kilotons of ordnance against cities. Raids on Dresden, Hamburg and Tokyo were as devastating as the atom bombs used against Hiroshima and Nagasaki', and there is an unbroken conceptual line from the strategic planners of the Second World War bombing offensives to the nuclear targeters of the early post-war years.

By 1948 the United States had an arsenal of about 50 atomic devices and was developing the means to deliver them over the Soviet Union by means of medium-range bombers which could be based in Europe. The first Soviet nuclear test in 1949 hastened the decision by the United States administration to proceed rapidly with the development of thermonuclear weapons and by 1953 both countries had tested such weapons.

The middle 1950s saw the acquisition, first by the United States and rather later by the Soviet Union, of large arsenals of high-yield thermonuclear weapons together with intercontinental-range bombers. The primary strategy was one of massive use of these weapons against the opponent's urban industrial base, with any inaccuracies in delivery largely countered by the massive multi-megaton yields of the weapons then available.[6]

[6]*Strategic Nuclear Targeting*, edited by Desmond Ball and Jeffrey Richelson (Ithaca and London: Cornell University Press, 1986) is a comprehensive account of the evolution of nuclear targeting.

Massive retaliation was the basis of strategy, with the United States able to respond to conventional and eventually nuclear attacks with the effective destruction of the Soviet Union.

Britain presents a useful example of the development of high-yield weapons for massive attack. The British decision to develop nuclear weapons was taken in early 1947 and resulted in a programme leading to the testing of a fission device in October 1952. More significantly, this was followed by rapid progress towards a thermonuclear capability, achieved in terms of a tested device in 1957. By the end of the 1950s, the Royal Air Force had deployed over 100 medium-range 'V-bombers'[7] which were being progressively equipped with a one-megaton thermonuclear free-fall bomb for use against Moscow and major cities in the western USSR. To ensure penetration of air defences, the stand-off Blue Steel missile, also carrying a thermonuclear warhead and with a 160 km range, was deployed in the early 1960s.[8]

By the middle 1960s, the United States and the Soviet Union were developing and deploying a nuclear triad of bombers. ICBMs and SLBMs. The bombers and ICBMs were equipped with high-yield thermonuclear warheads ranging up to 25 megatons. SLBMs, with their much more limited throw-weight, carried much smaller weapons. In all cases, the low levels of accuracy were such that only large industrial and urban targets were broadly suitable for attack, although large military complexes could also be targeted.

The strategic nuclear arsenals were seen as weapons of last resort but from the early 1950s onwards a wide range of much smaller-yield tactical nuclear weapons was being developed, these being seen as essentially war-fighting weapons. Nuclear-capable artillery was deployed by the US Army in Europe in the late 1950s and during the 1960s many other types of tactical nuclear weapon were developed and deployed. They included short-range unguided artillery missiles such as the Honest John, surface-to-air missiles such as the Nike Hercules and the SA-5 Ganef, many different types of low-yield fission bombs for delivery by medium and short-range strike aircraft, torpedoes, mines and even air-to-air missiles such as the Genie. All were integrated into the armed forces of the United States and the Soviet Union, and by the end of the 1960s, while on a smaller scale, this was also the case with Britain and France.

Thus, while strategic nuclear weapons were very high-yield devices suited to massive retaliation and assured destruction, at the tactical level, strategies for using low-yield nuclear weapons in limited wars were evolved.[9]

---

[7]Lorna Arnold's revealing account of the UK nuclear test programme in the 1950s, *A Very Special Relationship* (London: I1MSO, 1987) gives a valuable insight into the motives for Britain's development of nuclear weapons. Stewart Menaul's *Countdown: Britain's Strategic Nuclear Forces* (London: Robert Hale, 1980) is also interesting.

[8]See Robert Jackson, *Avro Vulcan* (Cambridge: Patrick Stephens, 1984).

[9]A detailed description of the various tactical nuclear weapons is available in Paul Rogers, *Guide to Nuclear Weapons, 1984–85* (Bradford: University of Bradford School of Peace Studies, 1984).

The late 1960s and early 1970s were also the years when early nuclear-armed ballistic missile defences were developed, notably the short-lived Spartan/Sprint system in the United States and the Soviet Galosh system around Moscow. Neither of these systems was successful, although the Galosh system was eventually upgraded in the mid-1980s. Indeed the Spartan/Sprint system remained in service for only a few months before being dismantled as ineffective and excessively costly.

By the middle 1970s two other developments were taking place at the strategic level which were to have a major impact on strategy. One was the ability to put multiple warheads on missiles, with each warhead being capable of independent targeting. Multiple independently-targetable re-entry vehicles (MIRV) were deployed on missiles first by the United States and then by the Soviet Union.

The second development was a series of considerable improvements in missile guidance which allowed accuracies of 200 m CEP[10] to be achieved by the end of the 1970s. Such accuracies meant that MIRVed missiles were beginning to develop the capability to destroy the opponent's land-based strategic nuclear forces, whether these were missiles in heavily protected underground silos or bombers protected within hardened aircraft shelters.

The first missile with counter-silo potential was the US Minuteman 3 ICBM, 300 of which were modified from 1980 onwards by having the NS-20 guidance system fitted together with the W78 warhead. This, rated at 335 kilotons, was nearly double the old W62 warhead. In the early 1980s; new versions of the Soviet SS-18 ICBM, while not as accurate as the Minuteman 3, had some counter-silo potential as they were fitted with a 500 kiloton warhead.

The development of highly accurate strategic counterforce[11] nuclear weapons accelerated during the 1980s. The MX 'Peacekeeper' ICBM entered service with the USAF in December 1986, with an accuracy twice that of the Minuteman 3, and in early 1987 there were reports of the Soviet Union testing an advanced and highly accurate version of the SS-18. The Soviet Union was also on the verge of deploying the new SS-25 ICBM, also likely to have counter-silo potential.

In one of the most impressive missile technology breakthroughs of all the United States started testing the Trident D5 submarine-launched ballistic missile in February 1987 and was due to start deploying it in 1989. The Trident D5 was a remarkable achievement, an SLBM sufficiently accurate to be able to destroy hardened targets. Previously, all SLBMs were far too inaccurate, largely due to their being launched from mobile platforms.

The significance of this overall trend towards counterforce strategic nuclear weapons was that, as at the tactical level, weapons were being developed which were more useable than the massive countervalue weapons of earlier years. If an opponent's nuclear arsenals were vulnerable to a surprise attack, pressures would exist under crisis conditions to undertake just such an attack.

---

[10]Rogers, *Guide to Nuclear Weapons*.

[11]Counterforce', meaning weapons intended for use against military targets as distinct from 'countervalue' weapons aimed at urban indus-trial targets.

There are powerful arguments against such a line of reasoning. It would be highly improbable that one side could ever destroy all the land-based missiles of its opponent, and even more unlikely that submarine-launched ballistic missiles could be eliminated. Even an attack limited to nuclear arsenals as distinct from the urban industrial centres would cause huge casualties, or 'collateral damage', likely to ensure that the country attacked would throw its residue of surviving weapons at the attacker.[12]

The merits of the different aspects of the first-strike argument are less necessary to analyse here than the appreciation that they form part of a trend, and that is the continuing efforts to overcome the main military disadvantage of nuclear weapons, that they are crude devices of mass destruction. The more accurate a nuclear weapon can be, the smaller need be the actual warhead required to destroy a specific military target, and the smaller the warhead then the lesser the collateral damage. As nuclear weapons get more precise, so they become more militarily useable. They approximate increasingly closely to the ultimate in military weapons, devices capable of delivering large bursts of energy very accurately over long distances at great speed. It is apparent that new developments are taking place in nuclear warhead technology which represent a new generation in weapons design and, in some cases, link in directly with the directed energy systems being developed within the Strategic Defence Initiative.

6.4 Three Generations of Nuclear Weapons

The fission devices or atom bombs developed in the early 1940s and used at Hiroshima and Nagasaki represented the first generation of nuclear weapons. In the early 1950s, second generation weapons, the so-called thermonuclear or fusion bombs were developed. All current strategic warheads and many tactical warheads are of this type.[13]

A key feature of all these first- and second-generation weapons is that their energy release is random in the sense that, at the moment of detonation, it is released in all directions. They might therefore be likened to a straight-forward high-explosive bomb in this respect but unlike a modern 'shaped-charge' explosive, such as is used in demolition work or in anti-armour weapons, in which the energy release may be sent in a particular direction.

[12]The counter-arguments are put in Malcolm Dando and Paul Rogers, *The Death of Deterrence* (London: CND Publications, 1984).

[13]First-generation nuclear weapons rely on the nuclear fision of uranium-235 or plutonium-239 for their energy release. In second generation weapons, the fission process is used to produce an intense burst of energy sufficient to trigger the fusion of a hydrogen isotope such as deuterium or tritium. This in turn is used to trigger a further fission process in the normally stable uranium-238, a fission fusion fission process, in other words.

This is not to say that nuclear warhead technology is crude. In the past thirty years there have been remarkable developments in the miniaturising of nuclear warheads, and recent designs have become highly specialised. Thus, one technology which may be applied to the task of destroying a silo is to use an earth-penetrating warhead[14] which penetrates into the ground at very high speed before detonating and producing an intense localised earthquake effect which disrupts a hardened silo. In an SDI environment, this can be married to a manoeuvrable re-entry vehicle designed to evade defensive missiles.[15]

Third-generation nuclear weapons involve a quite different approach in that they involve directed energy release. Either one component of the energy release of a weapon is accentuated, or else attempts are made to direct the energy release rather than allow it to be random. An example of the first is the enhanced radiation weapon (popularly called the 'neutron bomb') in which the neutron detonation is enhanced. An example of the second is the 'Excalibur' X-ray pumped laser being developed as part of the SDI, in which a small nuclear device emits intense X-ray bursts in particular directions.

Enhanced radiation weapons were developed by the United States in the late 1970s and two types were produced in the early 1980s, warheads lor the Lance tactical missile and the M110 203 mm self-propelled howitzers.[16] About 600 of these warheads have been stored at the US Army Arsenal at Seneca in New York State, specifically for deployment in time of conflict to Europe where their delivery systems arc already based. Public opposition to neutron weapons has made it politically difficult to deploy them routinely to Europe. France has developed neutron weapons but it appears that full-scale production was not planned until the late 1980s.[17] The most likely system to be fitted with a neutron warhead is the Hades tactical missile due to be deployed in 1992. US sources have occasionally suggested that the Soviet Union is developing neutron warheads, but Soviet sources deny this and there is no conclusive evidence suggesting current production by the Soviet Union.

Since the early 1980s, the two major nuclear weapons laboratories in the United States, Los Alamos and Lawrence Livermore, have undergone a transformation, as research on first- and second-generation nuclear weapons has been partially superseded by intense research and development into third-generation devices. Very little information has so far appeared in the open literature, but the technologies involve channelling the phenomenal force of a nuclear detonation into

---

[14]*Energy and Technology Review* for June July 1986 from the Lawrence Livermore Laboratory describes the laboratory's work on earth-penetrating warheads.

[15]It was reported in February 1987 that the USAF Ballistic Missile Office was investigating the use of manoeuvrable re-entry vehicles carrying earth-penetrating warheads for destroying silos in a strategic defence environment.

[16]See Dan Plesch, *Preparing for the First Use of Nuclear Weapons in Europe*, Peace Research Report Number 16 (Bradford: University of Bradford School of Peace Studies, 1987).

[17]See Paul Rogers, *Guide to Nuclear Weapons, 1987–88* (Leamington, UK: Berg Press, forthcoming).

some kind of destructive directed energy such as lasers, microwaves or even clouds of hypervelocity particles. While much of the work is currently SDI-related, it has considerable potential for many forms of offensive weapons, as will be discussed later.

While there are few indications that the Soviet Union is pursuing a similar path,[18] the US Department of Defense regards Soviet nuclear capabilities as broadly similar in technical level to those of the United States. This may be a worst-case assessment, but would suggest that the Soviet Union *does* have research and development programmes under way in directed energy nuclear devices.

The funding involved in the United States is both considerable and increasing. In part this is due to the fact that nuclear weapons involving new technologies require much more intensive testing than relatively straightforward designs. The latter may require around ten tests during their development programme, whereas the former may require up to one hundred.

Nuclear weapons developments in the United States are funded primarily by the Department of Energy rather than the Department of Defense and the DoE budget request for research and development of nuclear weapons in Fiscal Year 1987 was $603 million, up from $288 million in 1986.

An interesting feature of the research and development of new types of nuclear weapons is the enthusiasm created in the laboratories themselves, a facet admirably illustrated in recent interviews with Lawrence Livermore Laboratory staff. Giving evidence to a Congressional Committee in 1985, the Lawrence Livermore Director, Roger Batzel, said:

> What the potential of these systems will turn out to be in terms of offense or defense is not clear at this juncture these areas are new, they are idea-rich. And we have no reason to believe that we have more than scratched the surface with respect to the potential of these technologies.[19]

This was echoed by the head of the laboratory's weapons research, George Miller:

> We have only begun to understand the implications of this type of weapon, this type of technology. Precisely how it will be used ultimately, I can't really predict, but fundamentally it is very very different from what we've looked at or thought about in the past.[20]

As Livermore weapons designer, Ray E. Kidder, put it:

> The emphasis on the directed-energy stuff has given the lab a new lease of life. All the other stuff was getting to be pretty routine.[21]

[18] *Soviet Military Power* (Washington: Department of Defense, 1987) places some emphasis on presumed Soviet capabilities but is not specific.

[19] *Washington Post*, 9 June 1986.

[20] Ibid.

[21] Ibid.

6.5 The Real Implications of the SDI

The Strategic Defence Initiative was conceived as a means of developing effective defences against ballistic missiles. The defence technologies involved are varied. Some may involve sophisticated variants of existing technologies such as the development of highly accurate conventionally-armed missiles, but many will concern directed energy systems. These in turn may depend on beam weapons, either generated in space or generated on ground and reflected through space, the aim being to destroy missiles or re-entry vehicles in transit. The weapons themselves will form just one part of complex battle-management systems.

The targets to be destroyed will be small, fast and possibly hardened. They will require large bursts of energy which will have to be generated, and it is probable that one of the most important sources will be the use of nuclear detonations.[22]

It can be argued that comprehensive SDI systems will never be deployed, financial and technical considerations ensuring that they do not become feasible. Any long-term improvement in East West relations would certainly encourage this eventuality. But it can be further argued that this is irrelevant to the real significance of SDI and the related work on directional nuclear devices.

The technical problem for SDI is to be able to deliver energy with great precision over great distances in order to destroy ballistic missiles or their re-entry vehicles. Any move in that direction has much wider implications as such a delivery system can be developed for quite different purposes, to destroy not ballistic missiles but any other target relevant to the war-aims of a belligerent.

A space-based directed energy weapon can be developed as an offensive system. It could be used with great precision against military targets such as barracks or naval bases. It could be directed at a chemical manufacturing plant, an oil refinery, a defence ministry or a centre of population.

Its great advantage is that it would not involve the risking of military personnel, so it would lend itself to use in conflicts where domestic political considerations made it advisable to avoid casualties. Since space-based directed energy systems might be vulnerable to disruption by other states with a capacity to intervene in space, it is possible that their offensive value would lie in their use against Third World countries. The military advantages in such a context would be considerable. A country able to mount such offensive systems would be in a position of inescapable military superiority without any risk to its own personnel.

Other implications for the current high-level funding of directed energy weapons could be in the field of land- or sea-based conventional warfare, but it is probable that the most important developments lie in research into producing directional nuclear weapons. Such weapons would overcome the central military drawback to nuclear weapons, their collateral effects.

---

[22]An interesting example of a nuclear-pumped device involving the projection of hyper-velocity particles is the so-called Project Prometheus (see *Washington Times*, 22 April 1987), although this is claimed to be a method of destroying decoys rather than actual warheads.

In summary then, if a weapon is a device for transmitting energy to disrupt a target, any heavy investment in the research and development of technologies designed to enhance that ability may have major long-term significance. The Strategic Defence Initiative and its Soviet counterpart may be seen simply as defensive systems. In reality they are concerned with technologies which can be markedly offensive and which could approach the concept of an ideal weapon more effectively than previous exercises in military technology.

Chapter 7
Alternative European Defence (1990)

7.1 Introduction

Concern over the risk and effects of nuclear conflict developed in Western Europe in the late 1950s and lead to a significant movement of public protest, including the Campaign for Nuclear Disarmament (CND) in Britain.[1] The campaigning was essentially oppositional in that the prime motive was opposition to nuclear weapons and strategy, with relatively little thought given to alternative ideas on security.

In the 1960s, with the advent of the Partial Test Ban Treaty (PTBT) and the early Strategic Arms Limitation Treaty (SALT) discussions under way, public concern over the nuclear issue declined rapidly. By the end of the decade and notwithstanding a continuing evolution in nuclear strategy and weapons, interest in nuclear disarmament and non-nuclear alternatives had virtually disappeared. This remained the case until the end of the 1970s and the development of the new cold war, although there were a few exceptions.

In I958 Commander Stephen King-Hall had published his *Defence in the Nuclear Age* and over the following twenty years, a small group of researchers continued work on alternatives to nuclear strategies. The orientation of most of the work was non-military as well as non-nuclear, with civilian resistance serving as a major focus. Adam Roberts' *The Strategy of Civilian Defence* (1967), *War Without Weapons* (1974) by Anders Boserup and Andrew Mack and a number of books by Gene Sharp represent the main fruits of this work.[2]

Towards the end of the 1970s, there was a re-awakening of public interest in the nuclear issue, occasioned largely by the onset of a new phase of the Cold War. The Soviet intervention in Afghanistan, the coming of the more hawkish Reagan administration, the development of new weapons such as cruise, Pershing and the

[1]This text was first published as: "Alternative Military Options in Europe", in *Alternatives in European Security*, edited by Michael Randle, Paul Rogers, 1990, pp. 31–44, Dartmouth Publishing. The editors retained the copyright and granted the permission to include this text.
[2]See in particular Gene Sharp's The *Politics of Nonviolent Action*, Porter Sargent, Boston 1973.

P. Rogers, *Paul Rogers: A Pioneer in Critical Security Analysis and Public Engagement*, Pioneers in Arts, Humanities, Science, Engineering, Practice 21,
https://doi.org/10.1007/978-3-319-95150-8_7

SS-20, and a general perception of an accelerating nuclear arms race, all served to rekindle interest in nuclear disarmament.

A consequence of this was the perception of a need to look at alternative non-nuclear defence strategies, and the early 1980s saw a body of new writing and research in this area, one of the most substantial examples being the first report of the UK Alternative Defence Commission in 1983, *Defence Without the Bomb*.[3]

There was some modest overlap of this thinking with ideas being developed within the defence establishments of some western countries, where a concern to accentuate conventional deterrence led to some interest in aspects of non-offensive defence, but, for the most part, alternative defence research was undertaken in the civilian sector.

Two broad approaches to non-nuclear defence strategies can be identified. One is to take individual countries, even if they are currently members of an alliance, and examine how they may move towards purely non-offensive strategies. This is an approach which frequently borrows from existing policies of neutral states such as Sweden and Switzerland and has been applied particularly to the UK. The second approach is concerned much more with existing nuclear alliances, especially the NATO alliance, and seeks a much wider move towards a non-nuclear alternative. In practice, this approach has been applied most often to NATO forces in Central Europe with the main focus being on West Germany. This chapter will briefly review the more significant examples of each approach.

There is, though, a broader historical context which has to be recognised in any such assessment. Most of the work on alternative military strategies completed in the 1980s, has been in the context of a prevailing paradigm in which the Warsaw Pact is seen as an aggressive and markedly superior military alliance, comfortably outnumbering NATO forces in most of the recognised indicators of military power.

They have been further constrained by another central idea within the prevailing paradigm, that of military balance being essential for stability, what Boserop terms the belief in the fallacy of symmetry:

> This belief can have catastrophic consequences, especially when military technology undergoes rapid change as is the case today. Forces which are particularly suited for attack but which are of little avail for defence, for example, forces which have weapons which are both threatening and vulnerable, invite pre-emptive attack and encourage early use. To pursue symmetry in such forces is not to promote stability but to invite disaster.[4]

The NATO view is that it is not just necessary to have forces which are broadly equal in number, they must be similar in composition, even if they include substantial offensive components. Furthermore, if each side is tending to overestimate the forces available to the other, there is a tendency towards mutual escalation.

Warsaw Pact military superiority and the need to achieve balance has influenced the approaches adopted by peace researchers, even if they themselves did not accept

---

[3]Alternative Defence Commission, *Defence Without The Bomb*, Taylor and Francis, London 1983.

[4]Anders Boserup, *Non-offensive Defence in Europe* in Derek Paul (ed) *Defending Europe: Options for Security*, Taylor and Francis, London 1983.

this paradigm.[5] One could argue that they were therefore trying to produce non-nuclear military strategies for circumstances which they did not believe to be true, leading to distorted and unnecessarily expensive defence postures.

The counter argument is that as the prevailing paradigm is one of military inferiority and a potentially offensive and dangerous Warsaw Pact alliance, any proposals for non-offensive non-nuclear alternatives have much more chance of being taken seriously by the current political and military establishments, if they are formulated on the basis of this view of East-West relations.

In practice, most alternative defence proposals have tended to operate within the paradigm, while frequently criticising it as being, at the least, out-dated.

In the late 1980s, a new situation has arisen following the rapid changes in Soviet policies consequent on the advent of the Gorbachev administration and culminating in President Gorbachev's UN speech in December I988. Soviet arms control policy has, in western public perception, been transformed. An entirely threatening military posture is seen to have given way to a new approach which seeks a rapid improvement in East-West relations. The new Soviet approach has included a much greater frequency of meetings with senior western politicians, a much more positive approach to arms control, resulting already in the INF treaty, and a much more open discussion of defence policy. As Gerald Holden indicates elsewhere in this volume, there are already signs of a Soviet move towards perceptibly less offensive military postures.

While this shift may be in its infancy, and may be related largely to a Soviet desire to re-direct spending away from the military to the civil sectors of the economy, it has important implications for western researchers into alternative defence policies. It should, in particular, make it possible to integrate non-offensive military postures into a wider arms control strategy encompassing a variety of negotiated changes in forces. Mutual and codified progress towards alternative non-offensive strategies may utilise many of the ideas developed under the different circumstances of a dangerous phase of the cold war, but such bilateral or multilateral progress could be much more sustainable and could itself further aid an improvement in East-West relations.

Thus the aim here will be to review some of the more significant ideas on non-nuclear military alternatives, accepting that these have been developed in a different and more negative international context to that which may hopefully prevail over the next few years. The value of these ideas will then be assessed in terms of the changed climate, with particular reference to whether they can be used positively in the interests of common security.

---

[5]See Andrew Kelly, *Not By Numbers Alone: Assessing the Military Balance*, University of Bradford Peace Studies Papers, Number 11, 1984, which questions prevailing estimates of Soviet superiority.

7.2 The Single Country Approach

Researchers in a number of countries within the NATO alliance have looked at the potential for their countries to adopt non-nuclear and non-offensive defence policies. Studies of the West German situation have inevitably involved wider NATO issues, but, for countries such as Denmark and the UK, the work has frequently assumed the possibility of such countries adopting a non-aligned or even neutral stance in international relations.

In such cases, researchers have examined the defence policies of existing neutral and non-aligned states to see if their experience was relevant. Such states have tended to adopt one or more of three different strategies, frontier defence, defence in depth or protracted guerilla warfare. In all cases, the aim is to threaten to exact such a price following an invasion by a foreign power, as to deter that power from engaging in such an attack. The three states most often examined are Sweden, Switzerland and Yugoslavia.[6]

Sweden has a policy of armed neutrality which is not intended to deter a complete military occupation, but rather the taking of sites of major strategic importance, including ports and air bases, during a major East-West conflict. It has avoided war for over 160 years but maintains state-of-the-art military forces based largely on indigenous defence industries.

The navy is concerned essentially with coastal defence, and places greatest reliance on fast attack craft, mine laying and countermeasure vessels, coastal patrol craft and submarines. The navy lacks a capacity to land substantial forces on the territories of neighbouring countries, but could provide defence against such actions by others. The Swedish air force has invested heavily in modern high-performance interceptors which have a limited ground-attack capability.

The air force posture is directed primarily towards air defence, with virtually no capacity to engage in offensive air attacks on other states. The aircraft are highly versatile and can operate away from air bases, using stretches of highway for take-off and landing. Appropriate logistical support for such operations is available.

The army was originally configured primarily for frontier defence but now concerns itself also with defence in depth. There is emphasis on anti-armour tactics and mobility, with provision for guerrilla action in the event of loss of territory. The army, more than the navy and air force, relies heavily on conscripts. Overall in the armed forces, about 75% of personnel are conscripts, with former conscripts retaining reserve commitments which include refresher courses. The peace-time establishment of around 65,000 can be enlarged to some 800,000 by a three day mobilisation. Sweden also maintains civil and economic defence programmes, the latter including stockpiling of strategic materials, food and fuel.

Switzerland has a broadly similar political stance to Sweden. Being entirely landlocked it would appear, at first sight, to be intrinsically more vulnerable, but the terrain allows for much more effective defence in depth if not frontier defence.

[6]The following section is covered in greater detail in *Defence Without the Bomb*, *op. cit.*

Switzerland has a tradition of the citizen army and relies less heavily on the most modern weapons. One in ten of the entire population can be mobilised within 48 h to give an adult male army of some 600,000, dispersed throughout the country. This is backed up by an extensive civil defence programme and the stockpiling of strategic resources. A small air force is trained particularly to operate against attacking forces within the Swiss mountain environment.

Where Sweden stresses mobility over a large land area with a low population density; Switzerland stresses fixed defences throughout the much more densely populated country, albeit backed up by mobile forces. The aim is to extract an unacceptably high price of an attacking force by defence in depth, in the final analysis, resistance would be centred on high altitude defences which would control major trans-European communications through the Alps.

If Sweden and Switzerland have defence policies resulting partly from their historical experience of successful war avoidance, that of Yugoslavia stems partly from its experience under German occupation during the Second World War and its fear of a Soviet invasion in the post-war years. While Sweden and Switzerland have focussed on the need for defence against the side-effects of a European war, Yugoslavia has, at times, feared direct attack.

Its defence policy has been one of defence in depth followed by the provision for protracted guerrilla conflict. Some 60% of the standing armed forces of around 250,000 are conscripts, and a reserve of 500,000 is maintained. In addition to this, though, there are parallel Territorial and Civil Defence Forces of 3 million and 2 million respectively. The air force is configured for air defence and short-range ground attack, and the navy is essentially a strong coastal protection force of fast attack aircraft, patrol craft, mine layers and mine hunters.

If attacked, the army would attempt to hold the frontiers pending mobilisation of reserves and the much larger Territorial Defence Force, the latter organised on a local basis. Facing a full-scale invasion, the army would eventually withdraw and link up with elements of the TDF to engage in a protracted campaign of in-depth resistance, this serving as the major deterrent to attack.

The common features of the Swedish, Swiss and Yugoslav defence postures are threefold. Firstly, they do not depend on trying to achieve any kind of balance of military forces against those of a potential attacker. Secondly, they do not expect to be able to 'win' a war in the conventional sense of defeating the armed forces of the attacking power, and thirdly, they seek war avoidance by threatening to extract an unacceptably high price of an attacker.

7.3 Applications to Britain

Studies in non-nuclear defence strategies for Britain can be based on two different assumptions. One is that the UK is not part of the NATO unified military command, and, if not non-aligned, is certainly pursuing a much more independent foreign policy. The other is that Britain remains fully integrated into NATO. Studies based on the latter assumption commonly form part of studies of the NATO/Warsaw Pact confrontation in Central Europe, whereas studies visualising a more independent Britain borrow heavily from the experience of neutral and non-aligned countries.

Within NATO, Britain adopts an atypical defence posture in that it maintains its own nuclear forces and still seeks to maintain a global role. Among the members of the NATO military command, only the US has a similar posture, albeit on a much larger scale. Britain's defence posture has five main components:

- nuclear forces capable of independent use but normally committed to NATO,
- the British Army on the Rhine, a 50,000 strong force forming part of NATO ground forces in West Germany, backed up by RAF Germany with ground attack and strike aircraft and interceptors,
- naval forces responsible for much of NATO's North East Atlantic, North Sea and Channel security,
- forces for the defence of the UK, and
- forces for out-of-area operations.

Out-of-area commitments are considerable and have grown in the 1980s after declining in the previous decade. They include major forces in the Falklands and the Persian Gulf, and smaller permanent commitments in places such as Gibraltar, the West Indies, Ascension Island, Cyprus and Hong Kong. Global naval deployments are practised regularly, and in recent years these have included an annual round-the-world exercise by a small carrier task group.

Proposals for a non-nuclear and non-offensive defence posture for Britain restrict these commitments essentially to the defence of the UK. In one assessment, the army would be around 80,000 with 50,000 reservists, barely half current levels, and would operate almost entirely within the UK, with an emphasis on mobility.[7] The air force would be essentially an air defence force with a limited ground attack capability, and the navy would lose its carriers and most of its larger destroyer and frigate escorts, while maintaining a force of smaller escorts, patrol submarines and an increased force of fast attack and patrol craft. Assessments such as this typically involve a defence expenditure of 40–60% of recent levels, and would result in a defence posture not greatly dissimilar to that of Sweden, but with less reliance on reservists and more emphasis on enhancing the advantages of being an island. Nuclear forces, BAOR, and a major out-of-area capability would be cut out.

Such a view represents a middle line between those who advocate a much greater emphasis on civilian resistance and those who support continued membership of NATO's unified military command with some movement, within NATO, towards non-offensive defence. Essentially there is therefore a continuum from current policy through to an entirely non-violent security policy. Certainly, the Alternative Defence Commission recognised this and advocated that Britain's aim should be to keep at least one step ahead of what it hoped would become

[7]See Dan Smith, *Non-Nuclear Military Options for Britain*, University of Bradford Peace Studies Papers, Number 6, 1982.

international trends. Britain would thus adopt a non-nuclear stance, concentrate on defensive forces and lead the way with conventional force reductions, all of this taking some years, with the role being that of a catalyst for the promotion not just of a non-nuclear but ultimately a largely demilitarised continent.

While such ideas may be of interest in Britain, particularly among peace campaigners, they may be of less direct importance for other NATO countries, who correctly regard Britain as less than entirely central to the alliance. Consequently, while alternative defence policies for Britain have been examined more closely than in most other countries, greater interest within the NATO countries is aroused by research on more broad non-offensive defence policies applicable to the alliance as a whole, especially in relation to the inner German border, near which so many of the ground and air forces are located.

7.4 Debates Within NATO

In essence, NATO policy for responding to conflict in Central Europe has been to aim for strong forward defences allowing time for mobilisation, with provision for first-use of tactical nuclear weapons in the event of major incursions by Warsaw Pact forces. This was effectively the policy followed from the development of the flexible response nuclear first-use policy in 1967–8; however the new cold war of the early 1980s encouraged a mix of more offensive strategies to be developed.

These were primarily in relation to US policy and forces, although they were paralleled by a wider NATO debate, and they involved US forces other than those committed to NATO. Within Central Europe, the US Army's Air Land Battle ideas involved much greater mobility of forward-based ground forces and their integration with air power, the mobility extending to possible major incursions into Warsaw Pact territory early in a conflict. Related concepts included Follow-on Force Attack (FOFA), concerned with early air and missile strikes at second echelon Warsaw Pact forces well within Warsaw Pact territory, and the more general notion of Deep Strike, this being deep interdiction, again very early in a conflict, of many Warsaw Pact assets, including air bases, missile bases, choke points and logistic support.

Outside of Europe, the United States was simultaneously engaged in a considerable enhancement of its force projection capabilities. During the 1950s this would include the assembling of two additional carrier battle groups to make fifteen in all, the return to service of four battleships, armed with their original heavy guns and also cruise missiles for anti-ship and land attack, major improvements in US Marine Corps equipment, the expansion of special forces and a comprehensive programme to improve logistic support for long distance force projection.

This was all part of the 're-arming America' policy of the early Reagan years, and included the establishment of an entirely new military command, US Central Command (CENTCOM) covering South-West Asia and North-East Africa, and the development of the US Navy's Maritime Strategy. Under Navy Secretary John

Lehman, the Maritime Strategy included the idea of 'taking the war to the Soviets' in the very early stages of an East-West conflict. This would include aggressive operations against Soviet naval assets, including missile submarines, in the North East Atlantic and North West Pacific, and even the use of marines in operations to occupy some areas of Soviet territory in Northern Europe and East Asia.

Thus, in the early 1980s, a wide range of offensive postures was being adopted by US military forces, some of it affecting NATO within which the US was the dominant member. Interestingly though, Britain was probably the only other major NATO member (excluding France which was outside the unified military command) to be strong in support for such policies. Most member states were not even prepared to meet the commitment of a 3% per annum increase in defence spending.

Within this context of a US-led move towards more offensive postures, there was a particularly significant nuclear element but, at the same time, pressure within some NATO circles mounted for a re-appraisal of the potential for a more defensive orientation of conventional forces.

The developments in NATO nuclear strategy comprised a firm commitment to early first use of nuclear weapons, the SACEUR, General Bernard Rogers, indicating the nature of his orders in an International Defence Review interview (February 1986) in the following way.

> Before you lose the cohesiveness of the alliance - that is, before you are subject to (conventional Soviet military) penetration on a fairly broad scale - you *will* request, not you may, but you *will* request the use of nuclear weapons' (*emphasis* in the original).[8]

In addition, NATO had been under pressure for some years from West Germany to relate its first use policy to an immediate use of nuclear weapons against Soviet territory, rather than limiting nuclear use to short-range systems close to the Inner German Border. The advent of ground-launched cruise missiles and Pershing 2 ballistic missiles provided the weapons suited to such a posture.

Countervailing pressures within NATO were concerned with improving conventional forces to enhance conventional deterrence and diminish the risk of resorting to nuclear war. There is an apparent contradiction here, in that one tendency, to improve conventional defences, appears to run counter to early first use of nuclear weapons which would not allow time for a purely conventional response to a Warsaw Pact attack. An early nuclear first use policy *might* be intended purely us a contribution to deterrence as a whole, but it involves detailed force planning, training, targeting and other activities; it thus provides a predisposition to nuclear escalation in practice, whatever the actual original purpose of the policy.

In practice, open literature studies of improving conventional deterrence such as the Report of the European Security Study, *Strengthening Conventional Deterrence in Europe*,[9] retain a commitment to nuclear first use. The ESECS study, published in 1983, concentrated on a number of areas of deficiency in NATO's conventional

---

[8]Quoted in *International Defence Review*, February 1986.

[9]European Security Study, *Strengthening Conventional Deterrence in Europe: Proposals for the 1980s*, Macmillan, London 1983.

forces. It called for improvements in target acquisition, counterbattery firepower, interdicting follow-on forces and airpower, improvements in NATO command, control and communications systems (C^3) and the improvement in means of disrupting Warsaw Pact (C^3). While the ESECS work placed some emphasis on forward barriers to provide choke points, and use of light-weight anti-armour weapons, this formed a small part of a study which included many recommendations for enhancing offensive forces.

7.5 Alternative Approaches

During the 1980s, a number of studies have been published on alternative defence postures for NATO forces west of the Inner German Border. They include the work of Afheldt, Unterseher, Boeker, Saperstein, Boserup, von Bülow and others.[10] There is considerable overlap in their views, with each giving emphasis to particular aspects of what is commonly called non-provocative or non-offensive defence.

Most of the work has been concerned with ground forces, including gun and rocket artillery and helicopters, with less emphasis on air forces; and force dispositions, equipment and tactics tend to be determined with reference to three parameters. These are the degree of concern with frontier or forward defence, the concentration on defence in depth, and the size of standing armed forces relative to reserves.

West German concerns with avoiding a major land war on their soil have accentuated NATO's commitment to forward defence, and some of the alternative studies have effectively accepted this prerequisite. Frontier defences are enhanced in two ways. Actual fortifications are improved, although these may include buried pipes filled with slurry explosives to create armoured vehicle barriers when required. These barriers are used both to hold up attacking forces and concentrate them in narrow areas of penetration. In both cases, the forces are subject to a fire barrier in which a variety of munitions are delivered over distances of 1–150 km, by means which include artillery, mortars, surface to surface missiles, bombs and air-delivered stand-off weapons. There is an emphasis on a wide variety of means of detection and target acquisition, many of them pre-placed, with mobile reserves available to provide back-up in the event of a breach of forward defences.

The defences will place heavy reliance on highly mobile weapons systems which cannot penetrate deep into the opponent's territory. Thus helicopters, dedicated medium-range anti-armour aircraft like the A-10, and especially STOVL aircraft such as the Harrier, are particularly appropriate. The numbers of defending troops in the forward zone would be relatively small, with emphasis on automated detection and weapons systems. The aim of those studies which emphasise forward defence is

[10]A useful summary of these views is given in the September 1988 special issue of *Bulletin of the Atomic Scientists*: A New European Defence.

not to provide a completely impenetrable barrier to attacking forces – all of the studies have a defence-in-depth back-up – but they do seek to extract a very high attrition rate of attacking forces. They are aided in this by an assumption that the adoption of such tactics, with less reliance on an offensive strategy, would allow NATO to re-direct much of its manpower and military spending towards defensive systems. This would hugely enhance NATO's existing strengths, providing a powerful deterrent to an attacker.

Within the alternative defence field, other analysts consider that a concentration on forward defence is dangerous, in that if a breakthrough can be achieved, available defence-in-depth forces may not be adequate. They therefore put much more emphasis on defence over a depth of fifty to one hundred kilometres. The defence in depth studies tend to emphasise large numbers of small but well-armed units, highly mobile but operating within a defined area to limit C^3I problems. The units are, at most, lightly armoured, but are equipped with anti-armour weapons. Some studies include tank forces, and others place reliance on artillery. All incorporate the use of close air support, with an emphasis on NATO's moving to robust and versatile anti-armour aircraft and helicopters, the former having S/VTOVL capabilities to avoid interdiction of bases.

Long-range interdiction of attacking forces is not sought on the grounds that this is overtly offensive and invites early pre-emption, but high reliance is placed on interceptors and surface-to-air missiles. Defence in depth utilises the fact that the attacker immediately faces problems of extended supply lines, with the defender's mobile self-contained anti-armour units able to operate well behind the forward edge of battle area (FEBA).

One version of defence in depth is known as attrition defence and is summarised by Saperstein:

> Thus, what is envisaged in an attrition defense is a large number of lightly armed men diffusely scattered over a familiar landscape, taking advantage of every previously known shelter, awaiting the attacking armoured column. Some will fire their precision guided missiles at long range at the fronts of the attacking vehicles. Others will wait until they are by-passed and then fire at the much more vulnerable sides and rear of the aggressor. After expending their missiles, the defenders might be expected to retreat, rearm and engage again; they might be expected to surrender; or they might stay low and hidden until the immediate attackers have passed and then emerge to harass their rear, with light personnel weapons, as organised units or guerrillas. The latter possibility should certainly enhance the deterrent aspects of the suggested attrition defense since no aggressor would want to occupy a territory filled with scattered, hidden, armed men and/or women.[11]

Such a posture clearly requires a very high level of motivation on the part of the forces engaged in this attrition defence. His version places reliance on what he terms latent forces, reserves who know an area very well, who may have served a period as regular troops, but are not professional soldiers. This brings in the third parameter, the ratio between regular and reserve forces. Some analysts argue that

---

[11]See Alvin VI Saperstein, *Non-provocative defense and disengagement zones in Defending Europe, op. cit.*

the more cost-effective ground force operating in Central Europe would be one with a relatively small regular force, certainly less than half the size of, for example, current NATO forces in West Germany, but with an overall posture involving a much greater commitment to equipping and maintaining the expertise of a large number of reservists. This is thus borrowing from the Swiss model, the thinking being that the financial savings consequent on a smaller standing army would allow for a high level of training and equipping of reserves.

In essence, therefore, these models for alternative defence postures in Central Europe all seek, through enhanced conventional defence, to provide a deterrent to an attacker. There is no presumption that a war can be won, in the sense that the attacker can be defeated and its own territory counter-attacked and occupied. The aim, instead, is to make it unacceptably costly for an attacker to consider mounting an offensive.

7.6 Criticisms and NATO Responses

There is now a fairly substantial body of work on alternative defence postures, but it is reasonable to say that there is little evidence of this affecting military thinking within NATO as a whole, in the armed forces of some countries, including Norway, Denmark and, most significantly, West Germany, there is evidence of a changing outlook, but this is not reflected in corporate NATO attitudes.

There are still major gaps in the work on alternative security. Most attention has been focussed on ground forces, too little work has been done on countering offensive air forces, and there is little analysis of the naval balance.

Furthermore, most of the work has been applied to Western Europe, with little or no attention paid to South-East Europe or Western Asia. Nor has there been sufficient attention paid to NATO out-of-area activities, or the non-NATO aspects of the US Maritime Strategy and its continuing quest for maintaining and enhancing global naval supremacy.

Against this, where the detailed work is available, it is of sufficient depth to warrant a NATO response. The inertia within NATO is clearly substantial, and a response is not yet forthcoming. NATO as a whole appears to be in a serious predicament in its current response to the rapid changes in Soviet arms control policy, and to the early signs, outlined by Gerald Holden, of changing ideas within the Soviet Union to its military posture.

NATO has remained within a remarkably stable paradigm of the Soviet threat for more than three decades, it is simultaneously having to face up to several different trends which should encourage re-appraisal. Within the West, alternative defence analysts are saying that non-offensive postures can be adopted which are cheaper and aid stability. They are further arguing that the Soviet threat has been greatly exaggerated. At the same time, the policies now being pursued by the Soviet Union make it far more difficult for NATO to argue the seriousness of the Soviet threat to its wider political constituency.

The response so far has been one of conservatism and caution. The common NATO approach is to give a guarded welcome to any arms control developments while arguing that NATO must 'wait and see'. Since the Warsaw Pact are the 'bad guys', they must make all the running, whereas NATO must be careful, suspicious and very slow to responds The fact that this may limit the speed of Warsaw Pact initiatives is a necessary risk.

7.7 Non-offensive Defence and Arms Control

If NATO were to adopt a much more positive attitude to recent Soviet changes, and to adopt some alternative defence ideas, what would be the most promising lines to pursue? Bilateral and unilateral initiatives should work in parallel, with the latter intended for bilateral codification. Progress should be sought in nuclear and conventional postures.

Bilateral nuclear agreements could commence with the so-called 'third zero' proposal which extends the INF agreement to cover all short-range surface-to-air missiles. A unilateral freeze on air-borne theatre nuclear weapons could be accompanied by an offer to develop a treaty involving total withdrawal of these systems. While the dual-capable nature of the delivery vehicles presents problems, the rapid progress in verification procedures agreed under the INF treaty suggest these are far less an obstacle than previously thought.

Unilateral withdrawal of all artillery-fired nuclear shells would be widely acceptable in Western Europe, especially West Germany, and could also be accompanied by proposals for a verified bilateral treaty. While progress on naval tactical nuclear weapons involves systems deployed outside of European waters, progress should be possible on verified agreements aiming to create nuclear free zones, including the Baltic, and eastern Mediterranean.

In relation to conventional military forces, declaratory policy and actual force postures should be altered in concert. Thus policies tending towards deep strike should be discarded in favour of non-offensive defence, and changes should be commenced in force structures.

In broad outline, the latter would include a number of changes in existing trends. There would be less emphasis on procurement of main battle tanks, self-propelled artillery and medium range interdictors, in favour of mobile lightly armoured anti-tank ground forces. Anti-tank helicopters would receive more emphasis, as would STOVL ground attack and interceptor aircraft, the latter at the expense of longer range conventional interdictors such as the Tornado, F-lll and F-15E.

At the same time, a number of moves would be made to accelerate progress in conventional arms control. The aim here would be two-fold. One would be to seek overall reductions in conventional forces, some of these being asymmetric, though this would affect areas where NATO, as well as the Warsaw Pact, has larger forces. Bilateral agreements favouring non-offensive defence would be sought. One would concern setting progressively restrictive limits on the size, composition and frequency of military exercises, with emphasis on increasing restrictions on tanks and larger artillery pieces. Another would concern the establishment of zones of withdrawal of heavy armour, these being established initially either side of the inner German frontier, from an initial 50 km depth through to 300 km+.

As agreements were achieved, some military spending, and especially the procurement of the more offensive systems, could be scaled down. The mixture of unilateral and bilateral actions would introduce a time of considerable change, and a degree of uncertainty. It could be argued that acquiring larger forces of particular types, albeit of a non-offensive nature, would go against the larger aim of scaled down forces, but given present circumstances and alliance attitudes, it may be a necessary pre-requisite to larger progress, by increasing alliance confidence in defensive abilities. As bilateral progress was made, however, agreed decreases in offensive forces could be followed by across-the-board decreases of all forces.

These particular proposals are indicative rather than precise, and are modest in the sense that they do not entail massive unilateral steps. Such steps could be entertained by an individual country, such as thc UK, and might stimulate a more positive commitment on the part of the NATO alliance as a whole. They are modest because of the recognition that NATO as an alliance appears remarkably reluctant to respond to current opportunities offered by the Gorbachev administration, but they are sufficiently wide-ranging to catalyse, in concert with the opportunities now provided by Moscow, a decade or so of rapid progress away from the current offensive military stand-off which still characterises Europe at the end of the 1980s.

Chapter 8
Learning from the Cold War (2000)

The Cold War lasted for 45 years until 1990 and, since its ending, two broad views of the confrontation have emerged.[1] The first, held by most of the strategic studies community, is that the Cold War ended with a clear and unequivocal victory for the United States and its NATO allies, with the Soviet bloc collapsing first into the Commonwealth of Independent States and a number of independent Eastern European countries, and subsequently into an even looser alliance, with the Russian Federation itself under threat of decline if not disintegration through rebellions in Chechnya and elsewhere.

A core part of this analysis is the belief that the Cold War's nuclear confrontation was an essential part of the process. Nuclear weapons kept the peace, and did so without there being any real risk of nuclear war. Thus nuclear deterrence provided an essential stability within which the much greater free market economic success of the western liberal democracies could ultimately lead to the downfall of a rigid and authoritarian centrally-planned economic ideology. Central to this process was the way in which the Soviet Union and its Warsaw Pact allies were forced into crippling defence budgets in a desperate attempt to maintain military parity with NATO.

This view of the Cold War, seeing it as a clear victory for the western alliance stemming from a necessary and essentially safe process of maintaining very large military forces, leads to a post-Cold War paradigm in which military security remains the ultimate guarantee of western well-being. It may be a much more uncertain and volatile world in the early twenty-first century, and this may well demand different kinds of military force. In essence, though, adequate and appropriate military forces saw the western alliance safely through the Cold War, and suitably organised forces will serve a similar purpose in the future.

---

[1]The text was first published as: "Learning from the Cold War". Chapter in *Losing Control: Global Security in the 21st Century*, Paul Rogers, pp. 11–38, Pluto Press, London, 2000. The original copyright was retained by the author and granted for this publication.

P. Rogers, *Paul Rogers: A Pioneer in Critical Security Analysis and Public Engagement*, Pioneers in Arts, Humanities, Science, Engineering, Practice 21,
https://doi.org/10.1007/978-3-319-95150-8_8

There is a very different way of looking at the Cold War, an analysis that suggests that it is a thoroughly inappropriate model for ensuring future security, and that the Cold War period was highly dangerous and inordinately wasteful.[2] Furthermore, it has resulted in a momentum in the development of a range of military technologies that has lasted well beyond the end of the Cold War itself and may ensure that future conflicts could be devastating for individual states and dangerous for international security as a whole. This chapter explores this alternative view, contrasting defence spending and the consumption of talent with key indicators of social expenditure as well as some of the after-effects of the Cold War as military technologies spread across the world.

It looks, in particular, at the dangers of the nuclear confrontation itself, not just the development of nuclear war-fighting ideas but also the false alarms and crises and the numerous nuclear accidents, some involving the loss of nuclear weapons. Much of the information was in the public domain during the Cold War, if not widely known, but some of it has only come to light more recently. Put together, they indicate a degree of danger during many of the Cold War years which turns out to be fully supportive of the much-derided warnings of anti-nuclear campaigners and peace researchers, so often dismissed as scare-mongering at the time.

The general public perception of the Cold War nuclear era in the west was of a persistent threat from the Soviet bloc that had to be matched by the United States and its NATO allies. In particular, there was an enduring acceptance of the need for a balance of terror, with mutually assured destruction ensuring a stable confrontation. An examination of the history of the nuclear arms race suggests otherwise, with a picture emerging of an action-reaction arms race as each side tried to gain an advantage over the other. Added to this was the persistent fear, on both sides, of a surprise attack. Much of the nuclear targeting and planning was about preparing for worst-case scenarios, pervaded by a 'get your retaliation in first' mentality.

Above all, the public perception that nuclear weapons were solely weapons of last resort, was frankly false – the declaratory statements of the superpowers that supported this view were simply not matched either by targeting policy or by the very weapons that were developed and deployed. Although the Cold War might now be over, at least in its extreme form, the United States, Russia and the middle ranking nuclear powers such as Britain and France all retain nuclear forces, show no signs whatsoever of rejecting their nuclear postures, and see uses for nuclear armaments in the decades to come. For this reason, if for no other, it is worth discussing the broad features of the Cold War nuclear arms race and of current knowledge about the crises and accidents of that era.

---

[2]See, for example: Richard Ned Lebow and Janice Gross Stein, *We All Lost the Cold War*, (Princeton: Princeton University Press, 1994).

8.1 Origins and Early Development of the Nuclear Arms Race

Nuclear weapons were originally developed as a result of the collaborative Manhattan Project in the United States, with the first atomic device tested on 16 July 1945 and atom bombs dropped on Hiroshima and Nagasaki on 6 and 9 August that year. The bombs were small by the standards of the later years of the Cold War. Both yielded less than 20 kilotons (20,000 tons of high explosive equivalent), but they destroyed two cities and killed over 150,000 people. As well as being a massive research and development effort, the Manhattan Project included an immediate production capability, and had Japan not surrendered in August 1945 as many as ten more atom bombs could have been produced within six months. Even with the ending of the war, one source estimates that the US already had six nuclear weapons by the end of the year.[3]

Immediately after the end of the war, the United States began to build up a substantial nuclear weapons industry and progress was so rapid that it had an arsenal of 50 atom bombs within three years. By 1948, tensions with the Soviet Union were rising as the Cold War took root, and the Soviet Union was putting massive resources into a crash programme to counter the US nuclear monopoly, seen as a fundamental threat to the Soviet state.[4] In a further complication, the McMahon Energy Act was passed in the United States, one of its provisions being the ending of nuclear co-operation with wartime allies such as Britain. This was a key factor in encouraging Britain, still under the illusion that it was a superpower, to start its own atom bomb programme.

In the immediate post-war years there were limited efforts to bring nuclear weapons under international control within the fledgling UN system, with the US-inspired Baruch Plan and its Soviet counterpart, the Gromyko Plan, both failing to make progress. Instead, the Soviets made rapid progress in their own nuclear programme, exploding an atom bomb in 1949, several years earlier than western analysts had expected.

By the early 1950s, a full-scale nuclear arms race was in progress between the US and the USSR. With Europe divided, Berlin a source of tension and the Korean War causing some three million deaths, each side sought to produce substantial arsenals of atom bombs and planes or missiles to carry them, and to develop the immensely more destructive thermonuclear or fusion bomb. The United States

---

[3]Robert S. Norris and William M. Arkin, 'Global Nuclear Stockpiles, 1945–1997', *The Bulletin of the Atomic Scientists,* November/December, 1997.

[4]For a concise assessment of the early development of strategic nuclear weapons, see: Norman Polmar, *Strategic Weapons, an Introduction,* (New York: Crane Russak, 1982). Other sources are: Thomas B. Cochran, William M. Arkin, Milton M. Hoenig, *U.S. Nuclear Forces and Capabilities,* (Cambridge, Mass., Ballinger, 1984); Thomas B. Cochran, William M. Arkin, Robert S. Norris and Jeffrey I. Sands, *Soviet Nuclear Weapons,* (New York: Harper and Row, 1989); Charles Hansen, *U.S. Nuclear Weapons – The Secret History,* (New York: Orion Books, 1988).

tested a device in 1952 with the USSR following a year later; both states going on to develop largely thermonuclear arsenals in the following decade.

The arms race had many other facets, one of which was that each country had historic technological advantages that the other sought to match. During the Second World War, the United States had fought long-range conflicts in Europe and the Pacific and had, as a consequence, substantial experience of building long-range bombers. The Soviet Union, in contrast, had a relatively weak air force but a powerful army with substantial experience in artillery rockets. As a result, the US concentrated on developing nuclear-armed bombers while the Soviets also went for missiles, but each saw the need to match the other in their particular area of expertise.

By the mid-1950s, the United States and the Soviet Union were engaged in a full-scale nuclear arms race comprising very powerful nuclear weapons delivered initially by bombers. Concern over their vulnerability to air defences led to intensive programmes to develop ballistic missiles, initially with medium range missiles based in Europe, and an attempted Soviet basing in Cuba that resulted in the 1962 crisis.

In 1957, the Soviet Union tested an SS-6 long-range intercontinental ballistic missile (ICBM) and also launched the first satellite, Sputnik. Both events caused consternation in the United States, but it was actually the US that deployed the first ICBM two years later. Indeed, a feature of the whole of the Cold War was the deployment of large but relatively crude missiles by the Soviet Union, contrasting with smaller, more reliable and accurate weapons by the United States.

As well as a wholesale and immensely costly arms race under way between the superpowers, others were following suit. The sheer cost and technological commitment involved could be staggering. After the decision of the Atlee government to develop the atom bomb, Britain invested considerable resources in a crash programme, testing its first atom bomb in 1952 and a fusion bomb five years later. Even as these programmes were under way, the fragmented British aircraft industry was competing to develop bombers and, in an extraordinary period in the 1950s, Britain was buying three different nuclear bombers – the Valiant, Vulcan and Victor V-bombers, was developing the air-launched Blue Steel cruise missile and was even trying to develop a medium range ballistic missile, the Blue Streak, and a supersonic tactical bomber, the TSR-2, both later cancelled.[5]

All this was in addition to a wide range of conventional weapons programmes as well as maintaining 800,000 people in the armed forces. Only after Suez did British defence policy begin to ‘cut its cloth’, but British defence spending and its commitment to a range of intensely costly weapons systems was to continue to absorb

[5]A detailed description of the development and deployment of RAF nuclear forces was prepared in the early 1990s and subsequently declassified and published: Humphrey Wynn, *RAF Nuclear Deterrent Forces*, (London: HMSO, 1994).

money and talent throughout the Cold War and beyond as the country persisted in attempts to 'punch above its weight'.[6]

As well as the British, the French and Chinese were actively pursuing a nuclear option during the 1950s. The French succeeded in their first test in 1960 and the Chinese followed in 1964. Israel started developing its own programme, with French aid, in the 1960s and probably had a limited capability by the end of the decade, India tested a device in 1974 and South African and Pakistani programmes followed shortly afterwards.

8.2 The Nuclear Arms Race – Acceleration or Control?

With the benefit of hindsight, the acceleration of the nuclear arms race in the 1960s seems extraordinary. Hiroshima and Nagasaki had shown that just two small crude atom bombs could each destroy a large modern city, yet the superpowers were already building first hundreds and then thousands of much more powerful bombs. Each side was locked into a mutually reinforcing competition in which it perceived itself faced with an implacable opponent determined to gain a strategic advantage.

During the late 1950s, both superpowers developed extremely powerful weapons, with the largest of the free-fall bombs and missile warheads having destructive yields of 10–25 megatons (up to 25 million tons of high explosive equivalent). Used against cities, these would be utterly devastating – a 25 megaton warhead detonated at an altitude of 100,000 feet would cause near total devastation over an area of 500 square miles and would start serious fires over an area of 2,000 square miles. A single warhead exploded over London or New York would destroy the entire city and its suburbs, killing many millions of people.

By the early 1960s both the Americans and the Soviets had ICBMs armed with thermonuclear warheads. The early missiles were based on land, were liquid-fuelled and were slow to prepare for launching, making them vulnerable to surprise attack. If a missile took several hours to prepare for launching, but had a flight-time of less than thirty minutes, there was a vulnerability which led on readily to a 'use them or lose them' outlook that was to permeate strategic thinking and nuclear planning for the rest of the Cold War (Fig. 8.1).

More immediately, other solutions to this problem of missile vulnerability were sought. One was to develop ICBMs with storable solid fuels that could be kept at more or less instant readiness for launch. A second was to place them in heavily protected underground silos which could survive almost anything short of a direct nuclear strike, and the third was to develop submarine-launched ballistic missiles

[6]Malcolm Chalmers, *Paying for Defence,* (London: Pluto Press, 1986).

Year	United States	Soviet Union
1945	6	0
1950	370	5
1955	3,050	200
1960	20,400	1,600
1965	32,100	6,100
1970	26,500	11,650
1975	27,200	19,400
1980	23,900	30,100
1985	23,500	39,200
1990	21,800	37,000
1995	14,100	27,000

Note: All figures rounded. From 1990 onwards, figures include warheads put into storage, with the great majority of Russian nuclear forces no longer deployed but not dismantled. If the nuclear weapons of Britain, France and China are included, the peak year for global nuclear arsenals was 1986 with just under 70,000 nuclear warheads available.

Fig. 8.1 Superpower Nuclear Stockpiles (Robert S. Norris and William M Arkin, *Global Nuclear Stockpiles*)

(SLBMs) which could be fired from large nuclear-powered submarines when submerged and which were, in theory, impossible to detect and destroy.[7]

Both the United States and the Soviet Union went on to develop so-called 'triads' of strategic nuclear forces based on ICBMs, SLBMs and heavy bombers, with the Soviet Union placing most emphasis on ICBMs. For each leg of the triad, there was offensive and defensive competition. Thus each side tried to develop superior missiles and planes, with constant competition and interminable warnings, especially from industry and their armed forces, that the other side was getting ahead. Similarly, there was intensive competition to develop anti-aircraft defences, anti-submarine warfare and even anti-missile missiles.

During the course of the 1960s, there were some efforts to develop methods for controlling the nuclear arms race, prompted partly by public opposition to nuclear weapons, partly by the dangers exposed by the Cuban missile crisis in 1962 and partly by some advisers and politicians, including many in non-nuclear countries, who were genuinely concerned with the costs and the dangers of the East-West confrontation. A Limited Test Ban Treaty was agreed in 1963 and the first Strategic

[7]Norman Polmar, *Strategic Nuclear Weapons.*

Arms Limitation Treaty (SALT I) was negotiated later in the decade, though this could do no more than set substantial upper limits on the numbers of missiles and bombers to be deployed. The first of the nuclear-free zone treaties were also concluded, one covering Latin America and another trying to prevent deployments in outer space.

Perhaps more significant was the Non-Proliferation Treaty that sought to place firm limits on the proliferation of nuclear weapons to new states. While supported by many non-nuclear states, it was controversial from the start. Some states, such as India, saw it as a means of maintaining a nuclear *status quo* for a few powerful states while preventing others from gaining similar power. Within the treaty there was a provision for existing nuclear states to sign and ratify it, but they were required to embark on nuclear disarmament. As the treaty put it:

> Each of the Parties to the Treaty undertakes to pursue negotiations in good faith on effective measures relating to cessation of the nuclear arms race at an early date and to nuclear disarmament, and on a treaty on general and complete disarmament under strict and effective international control.[8]

By the time the Treaty was signed in 1968 and entered into force in 1970, production lines for several types of strategic nuclear weapon were operating intensively in the United States and the Soviet Union, and total nuclear arsenals, strategic and tactical, were numbered in the tens of thousands. Much of this stemmed from a perceived need to have very large forces to ensure the survival of some part of the nuclear arsenal in time of war, but the extent of the arsenals amounted to a remarkable degree of 'overkill'. Given that the destruction of a handful of cities and industrial centres could devastate the economies of the United States or the Soviet Union, there was already a developing air of unreality, but this was to be carried to extremes with two new strategic developments under way by the early 1970s.

The first was the development of multiple warheads on each missile, a number of smaller warheads giving a wider 'spread' over a larger target and ensuring greater destruction. These multiple re-entry vehicles were then superseded by multiple independently-targetable re-entry vehicles (MIRVs) in which each warhead could be directed at a different target. The US Poseidon SLBM was an example of a MIRVed missile – carrying 10 warheads (14 over a reduced range), it could direct them at targets spread over an area of well over 20,000 square miles.

In one of the greatest ironies of the whole Cold War era, this MIRVing of missiles coincided with the second round of the SALT negotiations from 1972 to 1979, which set out to put upper limits on the number of warheads for each superpower. Over this precise period, as the arms controllers were trying to set

[8]For an analysis of the origins and development of the Nuclear Non-Proliferation Treaty, see: Paul Rogers and Malcolm Dando, *The Directory of Nuclear, Biological and Chemical Arms and Disarmament, 1990*, (London: Tri-Service Press, 1990).

limits, the armourers were pushing ahead with warhead production and MIRVed missiles – over the seven years of the talks, the US increased its strategic warheads from 5,700 to 9,000 and the Soviets from 2,100 to 5,000.

In parallel with this MIRVing was a remarkable increase in warhead accuracy. Early missiles were able to deliver large single warheads to within one or two miles of their target, but major improvements in guidance made it theoretically possible for warheads to be accurate enough to destroy missiles in hardened underground silos. By 1980, the United States had a missile, the Minuteman III, which had an accuracy of 600 feet CEP (circular error probable – a 50% chance of a warhead landing within this distance of the target). The Soviets followed suit – although their warheads were less accurate, they compensated for this by having much more massive missiles carrying more destructive warheads.[9]

Highly accurate missiles carrying multiple warheads could theoretically 'disarm' an opponent's land-based missiles while leaving the attacker with many missiles in reserve. Developments in anti-submarine warfare and air defences also suggested that other legs of the strategic triad might have some vulnerability, and the notion of a 'disarming first strike' acquired impetus, leading to the development of hair-trigger responses such as launch-on-warning. In such a scenario, a proportion of a state's nuclear arsenal could be launched in a pre-programmed sequence when warning of an incoming attack had been received but before missiles hit their targets. A variant was launch-under-attack, in which retaliation would be delayed until the first detonations of an incoming salvo of missiles, but both strategies had deeply worrying aspects because of the possible effects of false alarms.

By the early 1980s, the United States had over 10,000 strategic nuclear warheads and the Soviet Union about 8,700, and missile and bomber programmes then in progress indicated that the two states would have well over 30,000 strategic warheads by the early 1990s. In the eventuality, the peak was reached in the late 1980s, and improving relations, the ending of the Cold War, and the negotiations on a new set of Strategic Arms Reduction Treaties (START) began to have an effect, both on total strategic arsenals and also on the most destabilising strategic weapons, the accurate multi-warhead missiles.

In terms of numbers of weapons and sheer destructive power, the Cold War nuclear arms race was characterised, at the strategic level, by an absurd degree of overkill, far greater than could have been warranted by even the greatest commitment to mutually assured destruction. In the public mind, the nuclear aspects of the Cold War were all about such a balance of terror. In reality, the targeting policies of both the United States and the Soviet Union were much more about being prepared to fight a nuclear war and emerge victorious. This orientation towards nuclear war-fighting was true of strategic forces for much of the Cold War, as it was also for tactical nuclear weapons where both NATO and the Warsaw Pact maintained policies of first use of nuclear weapons.

[9]Examples were later variants of the SS-18 ICBM. The Mod 4 variant of this missile, operational during the mid-1980s, carried ten MIRV warheads, each rated at 500 kilotons.

8.3 Strategic Nuclear Posturing and War-fighting

If the public perception of nuclear weapons was to see them as ultimate deterrents, then the declaratory nuclear weapons postures of the superpowers systematically endorsed this perception. The difficulty was that declaratory policies and actual deployment policies were never the same, even as governments tended to promote the idea of mutually assured destruction in the public mind. This contrasted with military thinking, as demonstrated powerfully by an exchange between Senator Tower and General Jones, then Chairman of the US Joint Chiefs of Staff, at a Senate hearing in 1979:

> *Senator Tower:* General Jones, what is your opinion of the theory of mutual assured destruction?
>
> *General Jones:* I think it is a very dangerous strategy. It is not the strategy we are implementing today within the military but it is a dangerous strategy…
>
> *Senator Tower:* Your professional military judgement is that it is a dangerous strategy and it is not the one we should follow?
>
> *General Jones:* I do not subscribe to the idea that we ever had it as our basic strategy. I have been involved with strategic forces since the early 1950s. We have always targeted military targets. There has been a lot of discussion…about different strategies. We followed orders, but basically, the strategy stayed the same in the implementation of targeting.
>
> *Senator Tower:* Unfortunately I am not sure that your opinion was always shared by your civilian superiors.
>
> *General Jones:* I agree that there have been some, including some in government, who have felt that all we require is a mutual assured destruction capability. I am separating that from our targeting instructions in the field…[10]

There is now ample evidence from the literature on nuclear targeting that both the Soviet Union and the United States targeted their opponent's nuclear forces. General Jones simply did not accept that the basic US strategy had ever been to achieve stability by threatening only retaliation against Soviet cities as in the popular misconception of nuclear strategy being based on assured destruction. Even as far back as the 1950s, there are indications that US nuclear planners recognised the advantages to be gained from a first strike with nuclear weapons at a time of crisis, an outlook illustrated by a briefing from the then head of Strategic Air Command, General Curtis le May:

> *Q:* How do SAC's plans fit with the stated national policy that the US will never strike the first blow?
>
> *A:* I have heard this stated many times and it sounds very fine. However, it is not in keeping with United States history… I want to make it clear that I am not advocating a preventive

[10]Hearings on the *Military Implications of the Treaty on the Limitations of Strategic Offensive Arms and Protocol Thereto (SALT II Treaty)*, Committee on Armed Services, United States Senate, 95th Congress, 1st Session, 1979.

> war; however, I believe that if the US is pushed into a corner far enough we would not hesitate to strike first...[11]

From the late 1950s, the United States produced a series of integrated strategic nuclear targeting plans, designed to bring together the targeting undertaken by the three legs of the triad, the ICBM, SLBM and bomber forces. These were (and are) known as the Single Integrated Operational Plans (SIOPs), and, with the rapid expansion of strategic arsenals, reached their peak in the early 1980s at the height of the Cold War.

By 1982, Desmond Ball could write of SIOP-5:

> As a result of these developments, the US target plans for strategic nuclear war are now extremely comprehensive. The current version of SIOP-5 includes more than 40,000 potential target installations, as compared to about 25,000 in 1974.[12]

According to Ball, SIOP-5 had targets in four broad categories. *Soviet nuclear forces* included ICBM and intermediate range missiles and their launch facilities and command centres, nuclear weapons storage sites, airfields with nuclear-capable aircraft and ballistic missile submarine bases. There were approximately 2,000 targets in this category. *Conventional military forces* included barracks, naval bases, supply depots and conventional airfields and comprised 20,000 targets. *The Military and Political Leadership* included command bunkers, key communications and intelligence facilities and ran to around 3,000 targets. Finally, *economic and industrial targets* included war-supporting industries such as munitions and weapons factories, transport and energy facilities and industries that might contribute to economic recovery after a nuclear war, such as coal, steel and cement production facilities. This economic and industrial base included about 15,000 targets.

Furthermore, targeting plans such as SIOP-5 involved a number of levels of nuclear war-fighting, ranging from limited use of strategic nuclear weapons for specified actions through to all-out nuclear war – a central nuclear exchange. Thus, *regional nuclear options* might involve the destruction of leading elements of an attacking force, a form of strategic nuclear targeting that was very similar to NATO's policy for using tactical nuclear weapons in Europe. *Limited nuclear options* might involve a more substantial use of nuclear weapons leading to the selective destruction of fixed enemy military or industrial targets. This might amount to an attack falling short of general destruction but might induce a negotiated end to nuclear war-fighting short of a central nuclear exchange.

SIOP-5 also included *major attack options* meaning the substantial use of strategic nuclear forces against a wide range of Soviet targets, but it also included a series of possibilities termed *selective attack options*. Some of these give a clear

[11]Quoted in: D. A. Rosenberg, "A Smoking, Radiating Ruin at the end of Two Hours: Documents on American Plans for Nuclear War with the Soviet Union, 1954–55", *International Security*, **6**, 1982, pp. 3–38.

[12]D. Ball, *Targeting for Strategic Deterrence*, Adelphi Paper Number 185, IISS, London, 1983.

idea of the tendency to nuclear war-fighting. They included the targeting of Soviet facilities adjacent to allied states, one example up until the end of the 1970s being the destruction of Soviet military facilities such as army bases and airfields close to Iran, then an ally of the United States.[13] Another selective option, in the event of a war involving China against the Soviet Union, would have been destroying Soviet military forces close to China, an option known colloquially as 'kicking the back door in'.[14]

According to Ball, within each of these groups of options, there was a wide range of further plans, including 'with-holds' – targets not attacked – such as population centres and national command and control systems. Specifically avoiding the destruction of population centres *per se* ran in the face of the declaratory policy of mutually assured destruction, although any major attack on military and political targets would have led to mass civilian casualties. Avoiding the destruction of command and control systems made it more possible, at least to the nuclear planners, to envisage a nuclear war being controlled, with a negotiated end to the conflict. It is clear that SIOP-5 and the deployment policy which it embraced was firmly based on the idea that nuclear wars could be fought without assured destruction – that victory was possible.

Furthermore, the idea of first strike was certainly embraced in the 1980s, just as General le May had indicated in the 1950s. As Ball comments:

> Special categories of targets have also been delineated for pre-emptive attacks against the Soviet Union and for launch-on-warning (LOW) or launch under attack (LUA) scenarios in the event of unequivocal warning of Soviet attack.[15]

By the early 1980s, another element was intruding into nuclear strategy, the Strategic Defence Initiative (SDI – popularly called 'Star Wars') and its Soviet equivalent. There had been limited deployment of anti-missile defences in the previous two decades, but mutual agreement embodied in the Anti-Ballistic Missile Treaty had limited this to help prevent a further escalation of offensive forces. SDI made superficial sense in that it appeared to suggest that a state could be fully defended from incoming ballistic missiles. It was politically popular with the Reagan administration and also with defence manufacturers who could see the possibilities of substantially expanded research and development budgets, but expert opinion was, from the start, dubious that any missile defence system could be so effective as to provide full protection from thousands of warheads. The much more troubling aspect of SDI was the idea that it might be employed in conjunction with highly accurate MIRVed missiles. While an SDI system could not protect

[13]Ibid.

[14]D. Ball, personal information.

[15]D. Ball, *Targeting for Strategic Deterrence.*

against an all-out missile attack, it might offer significant protection against a residual nuclear attack coming from a state that had lost the great majority of its nuclear forces to a disarming first strike.[16]

8.4 Tactical Nuclear Weapons and First Use

Reviewing the development and deployment of strategic nuclear weapons overall, it is evident that an enduring feature has been a commitment to nuclear war-fighting. Although far less is known of Soviet targeting policy, the available evidence, coupled with the nature and extent of the strategic systems developed during the Cold War, indicates that the Soviet outlook had many similar features.[17] This was also the case with the development of tactical nuclear weapons, where NATO has maintained a long-time policy of flexible response, an anodyne term that embraces first use.

Although the early nuclear weapons were essentially strategic – intended for use against the core assets of an opposing state, the development of nuclear weapons intended for tactical use within particular war zones was an early feature of the East-West nuclear confrontation. By the late 1950s, both the United States and the Soviet Union were developing relatively low-yield free-fall bombs as well as early forms of nuclear-capable artillery. Over the next 25 years, a remarkable array of tactical nuclear weapons was developed and deployed, covering almost every type of military posture.

As well as free-fall bombs, short-range battlefield missiles were developed along with nuclear-tipped anti-aircraft missiles and several types of nuclear artillery and mortars. Nuclear land mines known as atomic demolition munitions were developed that could be emplaced to destroy major bridges or tunnels or even block mountain passes. At sea, submarines were equipped with nuclear-tipped torpedoes, surface ships carried anti-submarine nuclear depth bombs which could be delivered by missile or helicopter, and aircraft carriers could fly off strike aircraft carrying several kinds of nuclear bomb. There were even air-to-air missiles such as the US Genie, that were nuclear-armed.[18]

By the 1980s, there were around 20,000 tactical nuclear weapons deployed by the United States and the Soviet Union, based in more than 15 countries and on warships and submarines throughout the world. In the great majority of cases, the

---

[16]A useful source on the Strategic Defense Initiative is: Paul B. Stares, *Space Weapons and U.S. Strategy*, (London and Sydney: Croom Helm, 1985). A more critical assessment of SDI is: Rip Bulkeley and Graham Spinardi, *Space Weapons: Deterrence or Delusion*, (London: Polity Press, 1986).

[17]For a substantive analysis of Soviet strategy, see: David Holloway, *The Soviet Union and the Arms Race*, 2nd edition, (New Haven and London: Yale University Press, 1984).

[18]For a description of tactical nuclear weapons in Europe towards the end of the Cold War, see: Paul Rogers, *Guide to Nuclear Weapons*, (Oxford: Berg Press, 1988).

presumption was that if such weapons were used, they would not necessarily involve an escalation to an all-out nuclear war. In other words, nuclear war-fighting could be controlled. In Europe, perhaps the most tense region of the Cold War nuclear confrontation, both alliances had policies of the first use of nuclear weapons in response to conventional attack.

For NATO in the 1950s, prior to the Soviet Union's having developed a large arsenal of nuclear weapons, the posture was codified in a military document MC14/2, colloquially terms the 'trip-wire' posture. Any Soviet attack against NATO would be met with a massive nuclear retaliation, including the use of US strategic nuclear forces, and this assumed that the US could destroy the Soviet Union's nuclear forces and its wider military potential without suffering unacceptable damage itself.

By the early 1960s, the Soviet Union was developing many classes of tactical and strategic nuclear weapons, making it less vulnerable to a US nuclear attack. In such circumstances, MC14/2 became far less acceptable to western military planners who consequently sought to develop a more flexible nuclear posture for NATO. This became known as flexible response and involved the ability to respond to Soviet military actions with a wide range of military forces, but also with the provision that nuclear weapons could be used first in such a way as to force the Soviet Union to halt any aggression and withdraw. Once again, there was the belief that a nuclear war could be fought and won.[19]

The new flexible response doctrine was progressively accepted by NATO member states in 1967 and 1968 and was codified in MC14/3 entitled *Overall Strategic Concept for the Defence of the NATO Area.* It was a posture with one particular advantage for the United States in that it might avoid nuclear weapons being used against its own territory. A US Army colonel expressed this rather candidly at the time, writing that the strategy recognizes the need for a capability to cope with situations short of general nuclear war and undertakes to maintain a forward posture designed to keep such situations as far away from the United States as possible. [20]

Flexible response was to remain in operation for most of the last quarter century of the Cold War, including periods of considerable tension in the early 1980s. Operational plans for nuclear use were (and are) developed by the Nuclear Activities Branch of the Supreme Headquarters, Allied Powers in Europe (SHAPE) near Mons in Belgium, operating in conjunction with the US Joint Strategic Target Planning Staff responsible for the SIOP strategic nuclear posture from its base in Omaha, Nebraska.

By the early 1970s, flexible response was well established under the Nuclear Operations Plan which embraced two levels of the use of tactical nuclear weapons against Soviet forces, selective options and general response. Selective options

[19]A detailed analysis of NATO's nuclear posture is: Dan Charles, *Nuclear Planning in NATO*, (Cambridge, Mass: Ballinger, 1987).

[20]Walter Beinke, "Flexible Response in Perspective", *Military Review*, November 1968, p. 48.

involved a variety of plans, many of them assuming first use of nuclear weapons against Warsaw Pact conventional forces. At the smallest level, these could include up to five small air-burst nuclear detonations intended as warning shots to demonstrate NATO's intent.

At a rather higher level of use were the so-called pre-packaged options involving up to 100 nuclear weapons, the US Army Field Manual at the time defining a package thus:

> a group of nuclear weapons of specific yields for use in a specific area and within a limited time to support a specific tactical goal... Each package must contain nuclear weapons sufficient to alter the tactical situation decisively and to accomplish the mission.[21]

While these different levels of selective use were thought to be possible ways of winning a nuclear war, the possibility remained that this would fail, and a more general nuclear exchange would result. This was the second level of use of tactical nuclear weapons and was termed a general nuclear response in which NATO nuclear forces in Europe would be used on a massive scale along with US strategic forces.

Thus, by the end of the 1970s, NATO had developed a flexible response strategy which involved detailed planning for the selective first use of nuclear weapons in the belief that a limited nuclear war could be won. This was followed, in the early 1980s, by two further developments that made the strategy even more risky.

Since the formulation of MC14/3, the West Germans had been unhappy at the prospect of short-range nuclear weapons being used, since this would involve huge civilian casualties among their own people. They therefore argued within NATO nuclear planning circles that first use should involve the immediate selective targeting of places within the Soviet Union. This required highly accurate ballistic missiles with a very short flight time yet with a range sufficient to destroy key targets such as command centres within the Soviet Union. The terminally-guided Pershing 2 missile was the first to have this capability and was deployed in West Germany from late 1983. It was one of a new generation of mobile 'theatre nuclear forces' (TNF) that included US ground-launched cruise missiles and Soviet SS-20 missiles, but it caused considerable concern in the Soviet Union which did not have a similarly accurate missile. [22]

In addition to the deployment of missiles such as the Pershing 2, there was also evidence that NATO was moving to a policy of early first use of nuclear weapons, one indication of this coming eventually from a remarkably candid interview given by the NATO supreme commander, General Bernard Rogers, who said that his orders were:

---

[21] *Operations: FM 100-5*, US Dept. of the Army, 1982.

[22] See Rogers, *op.cit.*, for details of the terminal guidance system of the Pershing 2 missile and for a description of the deployment of the Pershing 2 and ground-launched cruise missiles in western Europe in the early 1980s, and of the Soviet SS-20 missile during the same period.

> Before you lose the cohesiveness of the alliance – that is, before you are subject to (conventional Soviet military) penetration on a fairly broad scale – you *will* request, not you may, but you *will* request the use of nuclear weapons… (emphasis in the original). [23]

In presenting its nuclear policy to the public, much of the NATO argument was based on the idea of countering overwhelming Soviet conventional forces with nuclear weapons. With the benefit of hindsight, and observing the rapid collapse of Soviet conventional strength at the end of the Cold War, it is arguable whether the massive but largely conscript armies of the Warsaw Pact would have presented the threat to NATO that was so often supposed. Furthermore, there is evidence that first use was more about an overall concept of winning a war then just compensating for any perceived conventional imbalances. This was demonstrated in a revealing presentation of NATO nuclear policy by the UK government:

> The fundamental objective of maintaining the capability for selective sub-strategic use of theatre weapons is political – to demonstrate in advance that NATO has the capability and will to use nuclear weapons in a deliberate, politically controlled way with the objective of restoring deterrence by inducing the aggressor to make the decision to terminate his aggression and withdraw. The role of TNF is not to compensate for any imbalance in conventional forces. The achievement of conventional parity could have very positive consequences for the Alliance's strategy of deterrence. But it would not, of itself, obviate the need for theatre nuclear forces. [24]

United States strategic policy, together with NATO nuclear policy, were both mirrored by the policies of the Soviet Union. Although the US tended to maintain a technical lead, especially in such crucial areas as missile accuracy, the Soviet Union established a massive nuclear weapons industry, producing numbers of nuclear warheads which, as became clear after the end of the Cold War, even exceeded the apparently inflated western intelligence estimates of the Cold War years. Moreover, Soviet strategic weapons were often much more powerful than their US counterparts, somewhat compensating for US technical superiority.

For much of the Cold War era, the Soviet Union insisted that it had a policy of no first use of nuclear weapons. In more recent years, there is ample evidence that this would not have not been the policy in practice. If an East-West crisis had spilled over into a war, then Soviet policy in Europe would have been to escalate rapidly, including the early use of nuclear weapons. Quite possibly this was because of the Soviet awareness of NATO's technical superiority, especially the deployment of highly accurate missiles capable of destroying Soviet command and communications systems. [25]

Overall, there was a dynamic of two superpowers entrenched in a nuclear arms race where the first use of nuclear weapons was considered a rational part of

---

[23]Interview with General Bernard Rogers, *International Defense Review*, February 1986.

[24]*Third Report of the Select Committee on Foreign Affairs*, House of Commons, London, 1987–88, p. 35.

[25]A relevant contemporary analysis is: Ilana Kass and Michael J. Deane, 'The Role of Nuclear Weapons in the Modern Theatre Battlefield: The Current Soviet View', *Comparative Strategy*, Vol. 4, No. 3, 1984.

military planning and where strategic nuclear developments were so destabilising that tactics such as launch-on-warning were considered necessary developments. Underlying this was a belief that a nuclear war could be fought and won.

Anti-nuclear campaigners and a few peace researchers and conflict analysts questioned such policies at the time, but they were roundly criticised as pro-Soviet defeatists, even when they pointed to the risks of nuclear accidents and crisis instability. Since the ending of the Cold War, however, their views have begun to resonate with those of some senior retired military and even a few former nuclear planners who are now able to recognise the extent of the dangers faced during the Cold War, dangers exemplified both by the experience of actual nuclear accidents and also of crises which now appear to have been far more dangerous than acknowledged in public at the time.

8.5 Nuclear Accidents

During the Cold War years there were more than forty accidents involving nuclear weapons or military nuclear reactors, some of which resulted in radioactive contamination and others in the loss of nuclear weapons. In addition, the United States, the Soviet Union and Britain all experienced major problems with their nuclear weapons industries. In Britain, for example, a serious fire occurred at the Windscale plutonium production plant in 1957, when one of the reactor piles caught fire and burned for three days, releasing radioactive iodine and polonium into the atmosphere. Close to half a million gallons of milk from dairy farms in the area were withdrawn from consumption and poured into local rivers and the Irish Sea. In the United States, there were protracted problems with nuclear weapons plants at Hanford River and Rocky Flats, and US intelligence reports indicated major problems of radioactive contamination in the Soviet Union stemming from production accidents.

The record on nuclear weapons accidents is far from complete, especially in the case of the Soviet Union where information relates almost entirely to naval accidents in international waters, but what is already known indicates a series of problems stretching over four decades or more. [26]

One of the earliest accidents involved a B-47 bomber on an overseas flight in March 1956, probably to a US base in Europe, that was lost with its crew and two nuclear capsules and never found. Later that year, another B-47 crashed into a nuclear weapons storage igloo at RAF Lakenheath in Suffolk, killing the four crew members. The plane was unarmed but the fire that followed enveloped the store,

[26]Unless otherwise stated, details of nuclear weapons accidents draw on three sources: M. Gaines, 'Nuclear Weapons: How Safe?', *Flight International,* 4 July, 1981; Shaun Gregory and Alistair Edwards, 'A Handbook of Nuclear Weapons Accidents', *Peace Research Reports Number 20,* (Bradford: University of Bradford, 1988); Jan S. Breemer, 'Soviet Submarine Accidents, Background and Chronology', *Navy International,* May, 1986.

which contained three B6 nuclear bombs. The bombs were burnt and damaged although, fortunately, the high explosive elements did not detonate. Eighteen months later, another B-47 jettisoned a nuclear bomb over the Atlantic coast of Georgia after a mid-air accident. The bomb was not recovered.

During the 1960s, the United States maintained a large force of B-52 strategic bombers, and two were involved in serious accidents within four days in January 1966. At Palomares in Spain, a B-52 collided with a tanker aircraft. One of its four nuclear weapons was recovered from land, and another after a 15-week search at sea, but the high explosive components of the other two detonated, leading to substantial radioactive contamination. Over 1,400 tons of soil and vegetation had later to be moved to a safe storage site in Texas. In the second accident a B-52 crashed while seven miles out from Thule Air Force Base in Greenland. All four H-bombs were destroyed in the crash and 1.5 million gallons of ice, snow and water had to be removed in the decontamination operation.

Much more recently, in June 1995, a B-52 practising for an air show at Fairchild Air Force Base near Spokane, Washington, crashed within 50 feet of a nuclear weapons storage bunker. Witnesses reported that the pilot appeared to throw the plane into a turn to avoid striking the weapons storage area. The pilot was one of the four crew members killed in the crash.[27]

Both the United States and the Soviet Union suffered a series of nuclear submarine accidents during the Cold War, some involving the loss of nuclear weapons. In April 1970, for example, a Soviet *November*-class twin-reactor submarine sunk about 170 miles south-west of Land's End after a reactor fire. This class of boat was known to carry tactical nuclear weapons and western attempts at salvage were hindered by a Soviet naval vessel which took up station over the wreck for many months.

There were several serious incidents involving submarine bases. In 1966, a radiation leak is reported to have occurred in a nuclear-powered submarine near Polyarny, close to the Northern Fleet submarine base. According to some reports, members of the repair team may have died of radiation sickness. Early in 1970, another incident took place when a large explosion wrecked part of the Gorki submarine yards leading to radioactive contamination of the Volga River and its Black Sea estuary.

More recently, during the 1980s, the Soviet Union experienced a further series of submarine accidents, the most serious being the loss of a *Yankee*-class ballistic missile submarine in the Atlantic in October 1986. The 8,000-ton submarine carried 16 SS-N-6 ballistic missiles, each probably equipped with two 500 kiloton nuclear warheads rather than the earlier one-megaton warheads which originally equipped

---

[27]'US Jet Crash Close to Nuclear Weapons Area' *Reuters News Reports*, (RN01W, 042695.466), 26 April 1995.

this missile. The submarine experienced a fire and the explosion of the propellant of one of the missiles, killing three crew and blasting a hole through the side of the submarine's hull. The fire was eventually extinguished two days later, and an attempt was made to take the submarine in tow. This failed and the boat was lost with all its 32 nuclear warheads still on board.

Among the most serious of the US submarine accidents was the loss of the *USS Scorpion* with all of her crew, 450 miles south-west of the Azores on 21 May 1968. This followed an accident believed to have been caused by the explosion of an accidentally armed non-nuclear torpedo, but the submarine is also known to have been armed with nuclear-tipped anti-submarine weapons, probably the UUM-44A SUBROC weapon which carried the W55 warhead with a 1–5 kiloton yield.

Perhaps the most remarkable nuclear weapon accident of the Cold War years took place on 19 September 1980, and involved a Titan II ICBM, a liquid-fuelled missile carrying a massive 9-megaton W53 thermonuclear warhead in a Mark 6 re-entry vehicle. During routine maintenance of the missile, a mechanic dropped a wrench down a silo which hit the side of the missile and ruptured a fuel tank. Release of the propellant led to a large explosion several hours later that killed two men. The force of the explosion ejected the re-entry vehicle with its thermonuclear warhead from the silo and threw it 200 feet. Although not reported in the press at the time, it was later confirmed that there had been contamination of the site.

A near-farcical event took place four years later at a Minuteman III ICBM silo at Warren Air Force Base near Cheyenne, Wyoming, when a computer system appeared to indicate that the missile was about to launch itself. Air force officials promptly parked an armoured car on top of the silo door so that the missile would be irreparably damaged if it started to launch. Officials later insisted that there was no risk of an accidental launch but could not explain why it was necessary to take this unusual precaution. [28]

Although many of the nuclear accidents took place in locations away from immediate zones of tension, there were exceptions. One example, which did not specifically involve a nuclear weapon, is salient for other reasons. On 25 May 1982, an RAF Phantom interceptor in North Germany accidentally fire a Sidewinder air-to-air missile which locked on to an RAF Jaguar nuclear-capable strike aircraft that was within range. The missile severely damaged the plane which then crashed, although the pilot ejected safely. That such an accident could happen caused grave concern, the more so as it took place at a time of considerable tension. At the time, Britain was involved in a bitter war with Argentina over the control of the Falkland/Malvinas islands, a major crisis was developing between Israel, Lebanon and Syria, and East-West tensions were rising because of the impending deployment of cruise and Pershing missiles in western Europe.

[28]'Vehicle Parked on Silo After Launch Signal', *Washington Post,* 29 October, 1987.

8.6 The Experience of Cold War Crises

As far as is known, none of the numerous accidents with nuclear weapons during the Cold War caused the detonation of a nuclear weapon, and protagonists of nuclear strategy cite this as an indication of the extent of the safety measures in place. At the same time, its is likely that a number of the known incidents, and possibly some incidents which are still classified, may have been more dangerous than has so far become apparent. There are indications of this in the writings of some analysts with intimate knowledge of events of the period. One of the leading US strategists of the Reagan era was Fred Ikle, a strong supporter of a vigorous US military posture at the time. He has recently felt able to indicate his concerns about the problems of that era in writing about the potential dangers of future nuclear proliferation:

> …despite the several accidents and mistakes that could have sparked a large-scale nuclear war (and whose horrid details are still largely shrouded in secrecy), the superpowers always stopped just short of the abyss. At each of these fateful moments, the world escaped nuclear holocaust – seemingly by accident.[29]

During the Cold War there were a number of crises or false alarms, the seriousness of which is now clear, three of them dating to the earliest years of the Cold War. The Korean War, from 1950 to 1953, was officially a UN operation, although the leadership was American and the great majority of the forces ranged against the North Koreans and Chinese were American and South Korean. The US gave consideration to both the tactical and strategic use of nuclear weapons – tactical against troop concentrations and transit routes, and strategic against troop assembly areas and industrials targets both in North Korea and China.[30]

President Truman contemplated the risk of an escalation to a nuclear war involving the Soviet Union, writing in his journal in January 1952 that ‘It means that Moscow, St. Petersburg, Vladivostok, Peking, Shanghai, Port Arthur, Dairen, Odessa, Stalingrad, and every manufacturing plant in China and the Soviet Union will be eliminated.’[31] Unlike the head of the forces in Korea, General MacArthur, Truman was ultimately against the use of nuclear weapons, partly because of the risk of escalation to war with the Soviet Union involving Soviet action against European targets, but also because it was a UN operation and would involve, in MacArthur’s estimation, up to 50 of the limited stock of 400 nuclear weapons available in the early part of the war.[32]

The Korean War ended in stalemate in 1953, but the Indo-China war between the French and the Viet-Minh was entering its final phase in early 1954, with the

[29]Fred Charles Ikle, ‘The Second Coming of the Nuclear Age’, *Foreign Affairs,* Vol. 75, No. 1, January/February 1996.

[30]Norman Polmar, ‘Strategic Nuclear Weapons’.

[31]Quoted in Norman Polmar, ‘Strategic Nuclear Weapons’.

[32]Norman Polmar, ‘Strategic Nuclear Weapons’.

French facing a potentially catastrophic defeat as their strategic fortress town of Dien Bien Phu came under attack. The French requested US aid, and Operation Vulture was planned which would involve conventional and then nuclear strikes by US bombers against Viet Minh positions around Dien Bien Phu. According to Polmar, President Eisenhower and the majority of the Chiefs of Staff were in favour of Operation Vulture, but there were divisions in Washington, and the British were also opposed. A key factor was the reluctance of the French to accept US direction of the war, and the plan was not implemented. Dien Bien Phu fell to the Viet Minh early in May 1954.[33]

Later that year, a crisis developed over the status of the offshore islands of Quemoy and Matsu, controlled by the nationalist Chinese government on Taiwan but close to the coast of mainland China. Again, use of tactical nuclear weapons was considered by the United States in the event of a Chinese invasion of the islands, but the crisis was resolved early in 1955 without the Chinese taking that action.

Nearly a decade later, possibly the most serious crisis of the Cold War developed. In the summer and early autumn of 1962, the Soviet Union began to deploy medium-range ballistic missiles to Cuba, capable of reaching a wide range of targets in the United States. Part of the motive was to counter the US ability to hit Soviet targets with Thor and Jupiter missiles based in Britain and Turkey, partly it was to counter the perceived strategic threat from US ICBMs, especially the new solid-fuelled Minuteman missiles that were just starting to be deployed. When the US became aware of the Soviet move in October, 1962, it considered it to be a grave threat to US security, and a naval quarantine was ordered to prevent further Soviet shipments to Cuba.

As the crisis deepened, the US began to organise for a possible air assault against Cuba followed by an invasion which would involve 180,000 troops. US intelligence reported that there were only 10,000 Soviet troops in Cuba and that they did not have tactical nuclear weapons. In both respects the assessments were wrong – the Soviet Union had 43,000 troops on the island alongside more than 250,000 Cuban troops. Furthermore, the Soviet Union already had 90 tactical nuclear warheads in Cuba, and as the crisis reached its peak, warheads were moved from storage sites to positions close to their delivery systems in anticipation of an invasion.

The US Secretary of Defense at the time, Robert McNamara, has recently reviewed these events:

> Clearly, there was a high risk that, in the face of a US attack – which many in the US government, military and civilian alike, were prepared to recommend to President Kennedy – the Soviet forces in Cuba would have decided to use their nuclear weapons rather than lose them. We need not speculate about what would have happened in that event. We can predict the results with certainty. Although a US invasion force would not have been equipped with tactical nuclear warheads – the President and I had specifically prohibited that – no one should believe that had American troops been attacked with such weapons,

[33]Ibid.

> the US would have refrained from a nuclear response. And where would it have ended? In utter disaster, not just for the Soviet Union, Cuba and the United States but for all nations across the globe that would have suffered from the fall-out of the nuclear exchange.[34]

One of the most worrying aspects of the Cuban Missile Crisis concerns the behaviour of some sectors of the US military at the time. As the crisis developed, Strategic Air Command secretly placed nuclear warheads on nine of the ten test ICBMs being held at Vandenberg Air Force Base in California. This was a procedure to give them an additional supply of ICBMs that were just coming into service. Vandenberg was the main base for testing these new missiles and a tenth missile, which was unarmed, was launched on a scheduled test flight. This launch apparently took no account of the likely effect on Soviet intelligence if it had become aware of the arming of the other missiles.[35]

At one point in the crisis, the North American Air Defence command system received data indicating that a missile had been launched from Cuba and was about to hit the city of Tampa in Florida. In reality, a test tape had been fed into the computer system. Perhaps the most remarkable event took place at Malmstrom Air Force Base in Montana at the height of the crisis. Minuteman missiles were being installed at the base and air force technicians broke the safety rules and jury-rigged some missiles so that they could, *in extremis,* launch them themselves without receiving orders to do so.[36]

Eleven years after the Cuban Missile Crisis, the Yom Kippur/Ramadan war between Israel and Egypt and Syria was an occasion when both tactical and strategic nuclear weapons could have become involved. By 1973, Israel had a small number of Jericho ramp-launched surface-to-surface missiles fitted with 20-kiloton nuclear warheads. Early in the conflict, when Israel was hard pressed by the Egyptian attack through the Bar Lev line across the Suez Canal and by the Syrian attack across the Golan heights, there was huge concern that Israel's own territory might be occupied by Arab forces.

In three days of urgent activity, Jericho missile warheads were prepared for possible use at the Dimona nuclear weapons plant near Beersheba. The United States learnt of this, obtaining confirmation by a reconnaissance flight of an SR-71 Blackbird spy-plane. There is also some evidence that the Soviet Union was prepared to provide Egypt with a balancing force of nuclear warheads for its Scud missiles – a freighter with such warheads aboard is reported to have left the Nikolaev Naval Base at Odessa on the Black Sea.[37]

---

[34]Robert S. McNamara, 'The Conference on Disarmament Should Focus on Steps to Move Towards a "Nuclear-Free World"', *Disarmament Diplomacy,* April, 1996.

[35]Scott D. Sagan, 'More Will be Worse', chapter in: Scott D. Sagan and Kenneth N. Waltz, *The Spread of Nuclear Weapons, A Debate',* (New York and London: W. W. Norton and Company, 1995).

[36]Ibid.

[37]Colonel William V. Kennedy *et al., The Intelligence War,* (London: Salamander, 1987).

In the event, Israel succeeded in holding Egyptian and Syrian forces by conventional means, and within two weeks of the start of the conflict was threatening the Egyptian Third Army with defeat, aided, in part, by substantial logistical support from the United States. In these circumstances there were fears that the Soviet Union might intervene to prevent an Egyptian defeat, precipitating a major regional war. As part of its response, the United States moved its nuclear forces to an unusually high state of alert around the world, from DefCon 5 (Defence Condition 5, normal peace-time operations) to DefCon 3. In doing so, the United States neglected to inform its allies for several hours, including states such as Britain which was host to substantial US nuclear forces.[38] Again, further escalation was avoided, not least as Washington put pressure on Israel to accept a cease-fire short of a potentially unstable victory over Egypt.

Potential for nuclear use was not restricted to the superpowers. Britain deployed nuclear-capable V-bombers to Malaysia in the early 1960s at the time of confrontation with Indonesia over Borneo. Twenty years later, in 1982, tactical nuclear weapons were deployed on Task Force ships during the Falklands War, and a Polaris strategic missile submarine was deployed away from its normal area of patrol in the North Atlantic to bring it within range of Argentina.[39]

One of the most indicative crises of the Cold War happened in Europe in the autumn of 1993 and did not enter the public domain until several years later.[40] From 1979 through to 1983, East/West relations deteriorated markedly. In part, this was in response to the Soviet invasion of Afghanistan, but it is also related to the election of the hawkish Reagan and Thatcher governments in Washington and London, and the development of a range of new nuclear weapons systems including the US Trident, Pershing 2 and cruise missiles and the Soviet SS-20 and SS-24 missiles.

According to the Soviet defector, Oleg Gordievsky, the near-moribund Soviet leadership, led by the ageing Brezhnev, had expected the incoming Reagan administration of 1981 to tone down its vigorously anti-Soviet rhetoric once in office, emulating his Republican predecessor, Richard Nixon. In the event, the reverse happened as the US nuclear modernisation programme accelerated, bringing in more accurate ballistic missiles as well as open discussions about countervailing (war-winning) strategies.

---

[38]Robert Jackson, *Strike Force – The USAF in Britain Since 1948,* (London: 1986).

[39]A number of Task Force ships left Britain carrying stocks of anti-submarine nuclear depth bombs. There was a subsequent controversy within the Ministry of Defence, and most of the nuclear weapons were transferred to a fleet auxiliary, *RFA Regent.* While this ship went on to the South Atlantic to support the Task Force ships, it was subsequently kept away from the war zone, unlike its sister ship, *RFA Resource.* There remains uncertainty over whether ships that diverted from NATO exercise *Spring Train,* including *HMS Sheffield,* retained their nuclear weapons during the subsequent war. For a discussion of this and related issues, see: Paul Rogers, 'Sub-Strategic Trident: A Slow Burning Fuse', *London Defence Studies, No. 34.* (London: Brassey's, 1996).

[40]For a detailed description of the *Able Archer* incident, see: Paul Rogers and Malcolom Dando, *A Violent Peace: Global Security After the Cold War,* (London: Brassey's 1992).

One Soviet response, code-named RYAN, was to require operatives in western countries to watch out for possible preparations for war, a process started in November 1981 and still in progress two years later. By the autumn of 1983, Brezhnev had died, and his successor, Andropov was already terminally ill, Reagan was fully engaged in 'evil empire' rhetoric, was fully committed to the new Strategic Defence Initiative and deployment of the highly accurate Pershing 2 missiles was just starting in West Germany. To make matters worse, a civilian Korean Airlines Boeing 747 was shot down by a Soviet fighter when it strayed into Soviet airspace in the Fat East.

Early in November 1983, NATO commenced a set of highly secret exercises to test the release plans for nuclear warheads, *Able Archer* being the first such exercise to follow the deployment of the Pershing 2 missiles. Warsaw Pact surveillance systems monitored this process and NATO systems, in turn, monitored Warsaw Pact activity. According to one account, it quickly became apparent that NATO was 'listening in' to a gathering crisis:

Instead of the normal monitoring to be expected from across the Iron Curtain, a sharp increase was registered in both volume and urgency of the Eastern Bloc traffic. The incredible seemed to be happening, namely that the Warsaw Pact suspected it might really be facing nuclear attack at any moment.[41]

Documents later released from Warsaw Pact military committees in East Germany later confirmed this view:

On November 9, KGB stations in Europe were warned that American bases had been put on alert. The KGB suspected that a NATO exercise, *Able Archer*, could be a full-scale nuclear assault.[42]

As NATO officials became aware of the unexpected effect of *Able Archer,* significant changes were made to such exercises and the whole process of testing nuclear tactics was modified to avoid any such misinterpretations happening again.

8.7 A Crisis out of Nowhere

Perhaps one of the most remarkable examples of a sudden and entirely unpredictable crisis occurred several years after the end of the Cold War. Early on the morning of 25 January, 1995, a Norwegian-US research team launched a large four-stage Black Brant XII rocket as part of a long-term programme to study the Northern Lights. The rocket was launched from an island off the North Coast of Norway, an area that had been very sensitive during the Cold War, given that it was

[41]Gordon Brook-Shepherd, 'When the World Almost Went to War', *Sunday Telegraph,* 16 October 1988.

[42]Ibid.

close to a possible launch zone for US submarine-launched ballistic missiles aimed at targets in the former Soviet Union.[43]

To avoid false alarms, such experiments were notified in advance to the relevant Russian authorities, and the Norwegian Foreign Ministry had sent a letter to them reporting an impending launch of a research rocket in late January or early February, depending on weather conditions. Probably as a result of the chaotic state of the Russian bureaucracy at the time, this message had not been received by Russian radar crews. Moreover, the Black Brant XII rocket was much larger than previous experiments and its four stages resembled the multiple stages of a US submarine-launched Trident missile.

The Trident D5 missile carries six substantial thermonuclear warheads and, to the radar operators, it was not possible to dismiss the idea that it might be part of a surprise attack on Russia. Within minutes, the alert had reached the highest levels in Russia and, possibly for the first time, President Yeltsin's 'nuclear briefcase' was activated. It would have been possible for a Trident missile launched from off the Norwegian coast to have delivered its warheads over Moscow within twenty minutes, giving Yeltsin very little time to decide whether to launch a retaliatory strike. In the event, the early warning system was able to detect, within those twenty minutes, that the Black Brant rocket was not heading for Russian territory, and the alert status was reduced.[44]

The idea that such a crisis could develop in a matter of minutes 'out of the blue' may seem far-fetched, but there are two features of this incident that offer a degree of explanation. The first is that, if a surprise nuclear attack on the Soviet Union, or its successor state, Russia, had ever been contemplated, then it would almost certainly have started with submarines launching ballistic missiles from Arctic seas towards the command and control centres in Moscow, with some of the warheads fused to detonate at high altitudes to produce a massive electro-magnetic pulse (EMP) to disrupt electronic equipment in and around the capital.

Another explanation for this sudden and dangerous alert lay with an event in 1987 when a young German, Mathias Rust, flew a light aircraft right across Russia to Moscow and landed in Red Square. Rust's ability to evade detection by Soviet air defences had had a profound effect on the system, making operatives highly sensitive to the need to avoid getting caught out. Thus, at several levels up the ladder of command, officers decided to play safe and pass on the alert rather than check in detail before doing so.

When they first detected the research rocket, the radar operators could be blamed for causing a false alarm if they passed on details of the launch, but they were concerned that the rocket might just have been part of a missile attack, and passed on responsibility to a higher level. The general on duty adopted a similar stance, not

[43]Bruce G Blair, Harold A Feiveson and Frank N von Hippel, "Taking Nuclear Weapons off Hair-Trigger Alert", *Scientific American*, November 1997, pp. 42–49.

[44]David Hoffman, "Doctrines of the Cold War Refuse to Die", *Washington Post*, 15 March 1998, p. A1.

least because of the slight possibility that the missile might be equipped with an EMP warhead intended for Moscow as a prelude to a more general attack.[45]

Once the warning of the rocket's trajectory had been passed on to the Russian command and control system, Kazbek, predetermined procedures came into operation, ending with alerts in the three nuclear briefcases held by the Russian President, Defence Minister and Chief of Staff. With the command and control system now operating in combat mode, the trajectory of the rocket was monitored as Yeltsin and his colleagues conferred. Only when it was concluded that the rocket was not a threat was the alert terminated, the whole episode lasting just a few minutes.

8.8 Resourcing the Cold War

The 45 year period of the Cold War was an era of uniquely high military expenditure, outside of a world war. The peak was reached in the late 1980s, at a global level of well over $1 trillion dollars at 2000 prices, with the great majority of the expenditure being within the industrialised countries of the North. At the peak of the Cold War, public spending on the military even exceeded public spending on health or education. At 1987 prices, expenditure on the military was $926 billion, of which 84% was spent in the developed world, the overwhelming majority of it by the NATO and Warsaw Pact alliances. In that same year, world public expenditure on health was $794 billion and on education it was $904 billion. Put another way, spending by NATO and the Warsaw Pact on the military was $745 billion, whereas public spending on health and education in the whole of the developing world was less than one-fifth of this at $143 billion.[46]

Perhaps a more graphic contrast is demonstrated by referring to one particular global problem – food supply. In the early 1970s, the world experienced a sudden and unexpected down-turn in food production, with the effects being serious food shortages in many third world countries. At its peak the World Food Crisis in 1974 put some 40 million people at risk of famine in 22 countries across Asia, Africa and Latin America. One major UN response was the 1974 World Food Congress in Rome and, partially as a result of that conference, emergency steps were taken to meet the immediate shortages.

The conference went further than this, setting out a 10-year programme of public support for tropical agricultural development designed to improve self-reliance in food production across the Third World and prevent further crises. The support necessary for such a programme involved a three-fold increase in spending on agricultural development to $5 billion a year for the ten years (at 1974 prices). This amounted to approximately 2 per cent of world military spending at that time. In the

[45]Ibid.

[46]Ruth Leger Sivard, *World Military and Social Expenditure, 1996,* (Washington DC., World Priorities Inc.,).

eventuality, there were modest increases in agricultural development funding that came nowhere near meeting the demands, and further famines and food shortages were experienced in the 1980s and 1990s.[47]

There are innumerable other examples that can be drawn from the financing of the Cold War confrontation, and there is a frequent counter-argument that if the military spending had not been so high, there was no guarantee that social expenditure would have been improved. While this may be true, the fact remains that the Cold War was an extraordinary diversion of resources to what, in other respects, was a systematically dangerous confrontation.

This diversion of resources extends also to research expenditure. Supporting data is available for the whole of the Cold War period, but it is appropriate to take the year 1982, when the final stages of the Cold War were under way but before the peak commitments of the mid-1980s. In that year, US government-funded research and development on the military was $22 billion, compared with $11.3 billion for all government-funded civil research. The contrasts between military and health research are even more extreme – taking the United States and the European community together, and providing a per capita measure, military research spending was four times that of health research spending.

8.9 Proxy Wars

Although NATO and the Warsaw Pact never engaged in direct military conflict, the whole of the Cold War period was marked by proxy wars, fought in most regions of the world, in which major conflicts were fuelled by Cold War tensions, were enhanced by massive arms commitments from the superpower alliances, and were immensely costly in human terms.

Many of the conflicts in Central America, Southern Africa, the Horn of Africa, Indo-China, Korea and Afghanistan had direct or indirect superpower involvement. 100,000 people died in El Salvador and nearly as many in Nicaragua, 750,000 died in Angola, 1 million in Mozambique and over 900,000 in the Horn of Africa. In Asia the losses were even greater – 1.5 million in Afghanistan, 2.3 million in Vietnam and 3 million in Korea. While the Soviet Union and the United States had troops directly involved in some of these wars, not least Afghanistan and Vietnam, the overwhelming majority of the casualties were local people, mostly civilians. Indirectly, the proxy wars of the Cold War era were responsible for the deaths of at least 10 million people.[48]

---

[47]For an account of the 1974 world food crisis and the subsequent work of the UN World Food Congress, see: Paul Rogers, *Food in Our Time, But Not Yet* (London: World Development Movement, 1985).

[48]Casualty figures for wars since 1945 are given in: Ruth Leger Sivard, *World Military and Social Expenditure,* and *The Military Balance, 1999–2000,* (London: International Institute of Strategic Studies, 1999).

8.10 Proliferation of Weapons

The Cold War era was also a period of rapid developments in military technology, with many of these developments proliferating then, and since, across the world. In the decade since the end of the Cold War, as much of the conventional military equipment has been scaled down, a result has been a 'cascading' of light arms into zones of conflict or even of civil unrest. This has been especially marked in South Asia and Africa, with semiautomatic rifles and other light arms available on the grey and black markets for a fraction of their original cost.

The Cold War was also marked by the development and proliferation of nuclear, chemical and biological weapons of mass destruction, ballistic missiles, area-impact munitions and anti-personnel landmines. Nuclear proliferation was, to an extent, under some control in the latter part of the Cold War. There was even some limited progress as Brazil and Argentina abandoned their nuclear ambitions, South Africa gave up its small arsenal, and Ukraine, Belarus and Kazakstan returned the Soviet nuclear arsenals to Russia. More recently, though the India and Pakistani nuclear ambitions have accelerated, and Iraq was clearly involved in substantial nuclear ambitions.

There is a similar picture for chemical and biological weapons. The United States and the Soviet Union had massive chemical arsenals by the late 1980s, certainly over 50,000 tons each. They, and France, began a progressive disarmament in the 1990s, but states such as Egypt, Israel, Iran, Iraq, Libya, North Korea and Syria were still involved, although Iraq's CW programme was heavily limited by the activities of the UN Special Commission (UNSCOM).

The United States and the Soviet Union were heavily involved in biological weapons research, development and stockpiling in the early Cold War years. Along with Britain, France and many other countries, they were then committed to the Biological and Toxin Weapons Convention after 1972, although there is now abundant evidence that the Soviets continued with a covert programme, not least because they did not believe that the US, with its impressive expertise in genetic engineering and biotechnology, had abandoned its own programme. In the 1990s, there are repeated assessments from western intelligence agencies that a significant number of states maintain active offensive BW programmes, especially in the Middle East and East Asia.

For the past twenty years or more, there has been a strong public perception that missiles and bombs are getting steadily more accurate, and are now able to destroy targets without causing other damage. While this has indeed been a trend, although with many serious errors in recent wars in the Balkans and Middle East, it has been accompanied by another far less known development, that of area-impact munitions. The crude napalm of the Korean and Vietnam Wars has long since been overtaken by far more sophisticated devices. Among these are fuel-air explosives, cluster munitions and multiple-launch rocket systems. Cluster bombs typically distribute hundreds of 'bomblets', each exploding into hundreds or even thousands of high velocity shrapnel fragments. A standard British cluster bomb produces nearly 300,000 shrapnel shards over an acre and a half, shredding anything exposed to it. Some fuel-air explosive arrays can be detonated to give an explosive effect as damaging as a small tactical nuclear weapon, and a single salvo of a modern multiple-launch rocket system can destroy a small town in less than a minute.[49]

These kinds of weapons were developed in the 1960s and 1970s, initially by the US and the Soviet Union, but they are now produced by a dozen or more countries and are becoming more commonly used in war. Their effect is to make any major conventional conflict very much more devastating, especially when used against unprotected civilians.

8.11 Lessons to Learn

While this chapter has looked at the history of the Cold War years, and especially the nuclear confrontation, there are conclusions to be drawn that have plenty of relevance for the future. The first is that the sheer intensity of the nuclear arms race was almost unbelievable. Rational politicians, military leaders, planners and strategists were willing parties to a process of overkill that reached to the level of a farce. They, or most of them, were also caught up in targeting strategies that went far beyond the declaratory policy of deterrence through the fear of mutually assured destruction, and embraced nuclear war-fighting at both tactical and strategic levels.[50]

This necessarily embraced issues of first use, damage limitation and 'getting your retaliation in first', a process which incorporated highly accurate long-range multiple warhead missiles and ballistic missile defence programmes such as SDI. In particular, the existence of destabilising first strike weapons greatly increased the risk of instability at a time of crisis, yet the paradigm or mind-set was so firmly entrenched that there were very few people in the entire system that were able to break out of it.

Furthermore, for much of the Cold War, proponents of nuclear strategy were at pains to point out that there were numerous safeguards built into nuclear systems and that crisis management, too, was robust and stable. Here again, practical experience suggests otherwise, with more and more information coming to light both about nuclear accidents and the dangers of a number of major crises. If such problems can occur in a relatively stable, though massive confrontation, then such

[49]For an account of the status of area-impact munitions at the end of the Cold War, see: Paul Rogers and Malcolm Dando, 'The Directory of Nuclear, Biological and Chemical Arms and Disarmament'. See also successive editions of *Jane's Weapons Systems,* (London, Jane's Publishing Company).

[50]One of the most informative essays on attitudes of nuclear planners during the Cold War was written by a former target selection analyst: Henry T Nash, 'The Bureaucratization of Homicide', *The Bulletin of the Atomic Scientists,* April 1980.

behaviour can also be expected in much less stable and complicated instances of international insecurity in the future.

The Cold War was also a long period of military innovation, much of it permeating through the world in the aftermath of the Cold War. It was an extraordinary waste of human and economic resources and it consumed some of the most impressive research talent of most industrial countries for two generations. Powerful modern states lead by intelligent leaders and advised by experienced strategists failed to break out of a rigid confrontation for nearly half a century. For each 'side', its own perceptions of its own security needs made it impossible to break out from a narrow view and recognise the overall wastage and dangers of the confrontation. If that paradigm can exist for a substantial part of the world for so long, then similarly simplistic paradigms can be assembled for explaining and controlling the disorderly world that has replaced the Cold War era.

Part IV
Selected Texts on British Defence Policy

Chapter 9
Sub-strategic Trident (1996)

On Sunday, 7 January, 1996, the second of the Royal Navy's *Vanguard*-class Trident ballistic missile submarines, *HMS Victorious*, left its base at Faslane on the Clyde and went to sea to commence its first patrol.[1] The UK government had earlier disclosed that *Victorious* would be the first missile submarine to be deployed with a new single-warhead version of the Trident D-5 missile, a 'sub-strategic' system which was available for limited nuclear use.[2]

By 1999, the Trident system will be Britain's only nuclear system and will be deployed for both strategic and sub-strategic purposes. Since Trident is intended to have a service life of about thirty years, it is of interest to pursue some of the aspects of this dual function, especially its possible use in scenarios which fall far short of global nuclear exchanges.

My interest in this subject stems partly from an after-dinner conversation with a senior serving admiral at a defence college three years ago. This was at a time when there was still an intensive internal debate over the future of Britain's tactical nuclear forces. The Royal Navy was already facing the withdrawal of its tactical nuclear systems, the army had given up the dual-control nuclear artillery and battlefield systems shared with the US Army, but the RAF was maintaining its force of WE-177 tactical nuclear bombs for delivery by Tornado strike aircraft and was still interested in pursuing a stand-off missile replacement.

I raised with the admiral the notion that, with the ending of the Cold War, it was difficult to see the purpose of maintaining the Trident programme, intended as it was for deterring a major confrontation with a heavily armed Warsaw Pact adversary. He strongly disagreed, and pointed out the great advantages that Trident would have over nuclear-armed Tornadoes in responding to a regional crisis.

---

[1]This text was first published as: Sub-Strategic Trident: A Slow Burning Fuse, *London Defence Papers Number 34*, (London: Brasseys for the Centre for Defence Studies, King's College, London, April 1996).

[2]David Fairhall, 'Nuclear Deterrent aimed at Third World Dictators', *The Guardian*, 9 January 1996.

P. Rogers, *Paul Rogers: A Pioneer in Critical Security Analysis and Public Engagement*, Pioneers in Arts, Humanities, Science, Engineering, Practice 21,
https://doi.org/10.1007/978-3-319-95150-8_9

He cited as an example a future confrontation with a nuclear-armed state in the Middle East, believing it to be eminently practicable to use a Trident missile with a single warhead to fire a low-yield demonstration shot or, if necessary, use a combination of missiles and warheads in a pre-emptive strike against an opponent's nuclear facilities. Such operations could readily be undertaken by a *Vanguard*-class boat on patrol in the Atlantic.

To use nuclear-armed Tornado strike aircraft would, he argued, be considerably more difficult. They would have to fly out of a regional base, most likely RAF Akrotiri in Cyprus, and this might not appeal greatly to the government of the country concerned. Furthermore, they would probably be operating in an intense air defence environment, there would be the risk of loss of nuclear armed aircraft to anti-aircraft fire, and a number of aircraft and weapons would therefore be required to complete the task.

The conversation raised a number of interesting questions and contrasted markedly with the more general public view that Trident was very much a 'weapon of last resort', yet the admiral's views had a salience in relation to developments in a number of countries, most particularly the United States and France. Unlike these countries, though, there has been very little debate in Britain over the role of UK nuclear forces in the post-Cold War world, and the present paper is an introductory contribution to the topic.

9.1 Introduction

In September 1995, France commenced a series of nuclear tests in the South Pacific, a decision that prompted world-wide protests, not just by anti-nuclear activists but also by governments. In part, the reaction arose over issues of environmental damage to the Mururoa Atoll and surrounding seas. There was also a regional reaction in a number of French Pacific territories which was primarily anticolonial.

More generally, though, the problem with the French tests, as well as those conducted by China, was the way they contrasted with the negotiations earlier in 1995 which had resulted in the extension of the Nuclear Non-Proliferation Treaty (NPT). The NPT Review conference had focused attention on prospects for controlling proliferation and, along with the prospects for a Comprehensive Test Ban Treaty (CTBT), gave the impression that the nuclear era really might be coming to an end. The French tests seemed to fly in the face of this notion, lending support to the less comfortable idea that the nuclear era might instead be undergoing a transition.

British government reaction to the French nuclear test programme was muted in the extreme, but this is not altogether surprising. The French needed to test in order to complete development of two new warheads and also refine the technology to enable them to simulate the testing of future nuclear warhead designs. The British, by contrast, had their new nuclear system, Trident, already fully tested. Unlike the

French, they did not need to draw attention to their evolving nuclear posture and were therefore happy to keep their political heads down.

In practice, both states have firm plans for maintaining nuclear forces for at least the next thirty years, but this paper is concerned principally with examining the context of Britain's Trident programme. It concentrates primarily on Trident's 'sub-strategic' role, a phrase which has only been employed systematically in the last couple of years and implies a range of uses which were traditionally described as tactical or theatre but stop short of massive nuclear exchanges involving the full firepower of the system.

The long-term commitment to nuclear forces is a characteristic of all five declared nuclear powers – the United States, Russia, France, Britain and China. In the case of the western powers, there are two main motivations in this drive to maintain and modernise nuclear arsenals. One, inevitably, is the nuclear complex itself, a community of industrial producers, researchers, strategists and military who are firm believers in the utility of nuclear weapons and who also, perhaps, have the subsidiary motivation of maintaining their own careers. The other is the rapidly developing perception that the rigid confrontation of the Cold War era has been replaced by a volatile and unpredictable world in which nuclear, chemical and especially biological weapons will proliferate, a trend made worse by the collapse of the Soviet Union and the leakage of knowledge and materials from a fractured and even chaotic defence complex.

Into this unstable and unpredictable world, Britain now seeks to deploy Trident as a multipurpose nuclear attack system. Within its 30-year deployment time, there is a widespread assumption that there will be a proliferation of states with weapons of mass destruction, that there may be regional crises in which some such states threaten perceived western interests, and that there may even be occasions in which Trident or other equivalent western systems may be used in limited nuclear conflicts.

Such ideas are far removed from the persistent public perception of nuclear weapons as ultimate weapons whose use would inevitably occasion a global catastrophe, but this is a perception somewhat removed from the realities of nuclear strategy, deployments and targeting. Sub-strategic Trident is concerned with limited nuclear use, but this idea itself has a long history within British nuclear strategy, and it is useful to place this new use of Trident in this wider context. In doing so, there are three main elements. One is the long-term British involvement in NATO's nuclear planning, especially those elements concerned with first-use of nuclear weapons and the possibilities of fighting, and winning, a limited nuclear war. The second is the planning for deployment and possible use of nuclear weapons in out-of-area operations in the 1950s and 1960s, a process which related particularly to British security interests in the Middle East and South East Asia. The third is indicative, if incomplete, evidence of British deployment of nuclear weapons in more recent regional crises, especially the Falklands War.

All three elements will be explored in following sections, and the paper will then discuss attitudes to the future role of nuclear weapons currently evolving in the United States, Russia and France, and how they relate to British thinking.

Finally, there will be a preliminary discussion of the implications of such attitudes for the control of weapons of mass destruction.

9.2 Britain and NATO Nuclear Planning

Analysis of British nuclear thinking has been dominated for most of the past forty years by a concentration on strategic nuclear policy, especially in relation to the possible independent use of nuclear weapons. Tactical nuclear policy, within the NATO context, has been a subsidiary matter, but is significant in helping to understand the evolving role of Trident, especially when one looks at NATO policy in the most intense period of Cold War rivalry in the early 1980s.

At the height of the Cold War at this time, NATO, like the Warsaw Pact, had several thousand nuclear weapons in Europe. In NATO's case, these included Cruise, Pershing II and Lance missiles, bombs, artillery shells and mines. Most were controlled and owned by the United States which allowed allies such as West Germany, Belgium, Holland, Italy and Britain to operate them under a dual control system. Britain also had its own nuclear weapons and France had quite separate nuclear forces which were not integrated into NATO planning.

The Polaris system was represented as Britain's independent nuclear deterrent, although it was routinely committed to NATO for theatre nuclear targeting. In addition to Polaris, Britain's tactical nuclear forces were many and varied.[3] They included US-controlled nuclear depth-bombs carried by Nimrod anti-submarine aircraft, and US-controlled 155 and 203 mm artillery shells and Lance missile warheads for BAOR units in West Germany. In addition, the RAF had well over 100 British-made WE 177 tactical nuclear bombs which were carried principally by Tornado, Jaguar and Buccaneer strike aircraft. Finally, the Royal Navy had WE 177 nuclear bombs for its Sea Harrier VSTOL aircraft operating from Invincible-class light aircraft carriers, and an anti-submarine variant of the WE 177 which could be carried by helicopters deployed on most of the navy's destroyers and frigates. With all these forces, Britain was therefore a significant actor in NATO's nuclear scheme of things, and was integrated closely into the NATO nuclear planning process.

In the early years, until the late 1960s, NATO's nuclear policy, codified in a document MC14/2, was known as the 'tripwire' policy. This envisaged a massive nuclear response to any initiation of war by the Soviet bloc. It was a policy which had been developed in the 1950s at a time when the United States had a huge

[3]A comprehensive account of the development of British nuclear systems is in Robert S. Norris, Andrew S. Burrows and Richard W. Fieldhouse, *Nuclear Weapons Databook: Volume 5, British French and Chinese Nuclear Weapons,* Westview Press, 1994.

nuclear superiority over the Soviets. Even then, though, the United States was leaning towards the use of tactical nuclear weapons in scenarios which might fall short of total war. This resulted in some tensions between US and UK strategists, with the latter wanting to downplay the possibility of fighting a limited nuclear war in Europe. The view of the UK Chiefs of Staff was that if tactical nuclear weapons were used in Europe, escalation was inevitable.[4] By the mid-1960s, though, the tripwire policy of MC14/2 itself became untenable as the Soviet Union developed its own wide-ranging tactical and theatre nuclear forces for deployment in Europe, complementing its strategic forces and matching NATO at each level of potential escalation.[5]

As a result, NATO's doctrine of 'flexible response' was introduced, codified in document MC14/3 of 16 January 1968. This was entitled the Overall Strategic Concept for the Defence of the NATO Area, and covered general conventional and nuclear policy, with the details of the latter developed by the Nuclear Activities Branch of the Supreme Headquarters, Allied Powers in Europe (SHAPE), at Mons in Belgium. This group developed the nuclear components of NATO policy, setting out plans for flexible response in a document entitled Concepts for the Role of Theatre Nuclear Strike Forces in Allied Command Europe, completed initially in 1970 but subsequently updated.

Flexible response envisaged two levels of nuclear employment, selective use and general response. The former involved the use of a limited number of nuclear weapons, mostly low-yield warheads, against Warsaw Pact troops and their immediate logistic support, in the belief that they could be 'stopped dead in their tracks' so to speak. Numbers of weapons might be limited to a handful, little more than demonstration shots, or might number up to one hundred depending on circumstances.[6]

This policy was not promoted widely in public, although there were occasions when it was made clear, if not shouted from the rooftops. One example was evidence given by the UK Ministry of Defence to a select committee of the House of Commons:

> The fundamental objective of maintaining the capability for selective sub-strategic use of theatre weapons is political – to demonstrate in advance that NATO has the capability and will to use nuclear weapons in a deliberate, politically-controlled way with the objective of restoring deterrence by inducing the aggressor to terminate his aggression and withdraw.

---

[4]John Baylis, *Ambiguity and Deterrence: British Nuclear Strategy 1945–1964*, Clarendon Press, Oxford, 1995, p. 334.

[5]Daniel Charles, *Nuclear Planning in NATO*, Ballinger Publishing Company, Cambridge, Massachussetts, 1987, pp. 12–16.

[6]Although dealing primarily with strategic targeting, a good source on NATO nuclear planning is: Desmond Ball, Targeting for Strategic Deterrence, *Adelphi Papers* Number 185, International Institute for Strategic Studies, London, 1983.

> The role of TNF [Theatre Nuclear Forces] is not to compensate for any imbalance in conventional forces. The achievement of conventional parity could have very positive consequences for the Alliance's strategy of deterrence. But it would not, of itself, obviate the need for theatre nuclear forces.[7]

Selective use would thus be employed in the belief that Warsaw Pact forces would cease their aggression and withdraw, but in the event of escalation, then NATO's second strand of nuclear policy, a more general nuclear response, would come into use. This would involve hundreds or even thousands of nuclear weapons being used against a wide range of targets in Eastern Europe and the western part of the Soviet Union. Such a scenario would involve coordinated action with US strategic nuclear weapons, in other words, general nuclear war.

In the years of heightened Cold War tension in the early 1980s, with the hawkish policies of President Reagan and Prime Minister Thatcher very much in the ascendency, two further developments caused concern. One was an indication that NATO's first-use of nuclear weapons would involve immediate demonstration shots, not just against Warsaw Pact forces engaged in conflict against NATO forces, but also against targets in the western Soviet Union itself, using the new and highly accurate Pershing II ballistic missile. This provoked fears that NATO first-use would quickly be seen by Soviet planners to threaten their core interests, with unpredictable effects.

The second development was increasing evidence that NATO was adopting more aggressive essentially conventional tactics, including concepts such as deep strike, follow-on forces attack and AirLand battle, all concerned in different ways with taking a war deep into Warsaw Pact territory at a very early stage in a conflict.

Added to this was an apparent commitment to early first use of nuclear weapons. This was demonstrated by the Supreme Allied Commander, Europe (SACEUR), General Bernard Rogers, describing the nature of his orders in an interview published early in 1986:

> Before you lose the cohesiveness of the alliance – that is, before you are subject to (conventional Soviet military) penetration on a fairly broad scale – you will request, not you may, but you *will* request the use of nuclear weapons'. (Emphasis in the original)[8]

Up until the mid-1970s, the Soviet Union's view was that a major East-West conflict would rapidly escalate to all-out war. This view changed, with a developing belief that conventional forces might be able to achieve objectives in the event of war in Europe without recourse to nuclear weapons.[9] The NATO deployments of the early 1980s, especially the capability of the Pershing 2 missile for accurate time-urgent targeting of command centres, was viewed with particular concern, and there was, once again, a ready realisation of the risk of rapid nuclear escalation.

---

[7]Third Report of the House of Commons Select Committee on Foreign Affairs 1987–88, p. 35, para 6.

[8]Published in *International Defence Review*, February 1986.

[9]Michael McGwire, *Perestroika and National Security*, The Brookings Institution, Washington DC, 1991, especially Chapter 3.

Opposing sides were thus locked into a somewhat unstable nuclear relationship. We also know that in November 1983, the Soviet Union misread a routine NATO nuclear planning exercise, 'Able Archer' as a preparation for war, and came close to panicking in a manner which caused NATO to rethink its nuclear exercise process.[10]

Incidentally, nuclear archaeology is also tending to show that a number of other events, including the Quemoy/Matsu crisis with China in the 1950s, the 1962 Cuban missile crisis, and the 1973 Yom Kippur/Ramadan War were all rather more serious nuclear crises than appeared at the time.[11] In the Cuban missile crisis, for example, it now appears that the Soviet Union already had some tactical nuclear weapons in Cuba before the crisis erupted, so an attempted US invasion could have resulted in nuclear use. Furthermore, during the crisis, US fighters were airborne from bases in Alaska equipped with Genie air-to-air nuclear-tipped missiles.

In nuclear terms, the Cold War was not as cold as it seemed, and people at many levels within the armed forces of the UK would have learnt to live with the possibility of nuclear use. This was coupled with a policy which envisaged fighting a limited nuclear war, accompanied by a complex set of nuclear deployments, planning and training which were all based on this policy. From the 1960s onwards, Britain was an integrated component of this entire process, and the NATO experience is one important part of the evolution of Britain's nuclear policy.

9.3 Some Aspects of British Out-of-Area Nuclear Deployments

Probably as important in the context of sub-strategic Trident as Britain's involvement in NATO nuclear war planning is the long history of deploying nuclear weapons outside Europe, especially in the Middle East and South East Asia, but also on Royal Navy warships.[12] For the most part, these deployments related to the broad East-West confrontation, but they extended also to regional crises, even some involving non-nuclear powers. They also had much to do with an enduring belief that nuclear weapons were just another category of weapon rather than something entirely different, a belief itself backed up by a deep feeling within the British defence establishment, especially in the early nuclear years of the 1950s and 1960s,

---

[10] I believe the first account of the remarkable incident was actually published before the end of the Cold War in an article by Gordon Brook-Shepherd in *The Sunday Telegraph,* 16 October 1988, alter he had interviewed the Soviet defector, Oleg Gordievsky. It is discussed in some detail by Malcolm Dando in: Paul Rogers and Malcolm Dando, *A Violent Peace, Global Security After the Cold War,* Brassey's, London, 1992.

[11] Sean M. Lynn-Jones, Steven E Miller and Stephen van Fvera (editors), *Nuclear Diplomacy and Crisis Management,* The MIT Press, Cambridge Massachusetts, 1990.

[12] See Norris et al., *op. cit.*

that Britain was one of the world's great states and could support this claim with its nuclear forces.

In the mid-1960s, there were regular detachments of V-bombers to the Far East Air Force base at RAF Tengah in Singapore, including detachments of nuclear-armed aircraft. From 1961 to 1969, nuclear-capable Canberra bombers and nuclear weapons were based at RAF Akrotiri in Cyprus in support of the Central Treaty Organisation (CENTO). They were replaced by Vulcan bombers which were based there until 1975.

Deployments in the Far East were of particular interest because they followed a period of debate in the 1950s in which a British role in a limited regional nuclear war was considered. According to a study by Martin Navias, the Chiefs of Staff Committee (COSC) believed that a conventional war could well break out in the Far East if, for example, China committed an aggressive act against SEATO or Hong Kong, or if a conflict escalated between China and the United States following a crisis over Formosa or Korea.

The Chiefs of Staff believed that such a war would be communist-instigated but that it could be controlled by the use of nuclear weapons. Significantly, they were adamant that nuclear weapons could be used without the risk of such a conflict escalating to the strategic level with exchanges of nuclear and thermonuclear weapons taking place between the superpowers, China, and Britain.

There was therefore concurrence in the COSC that nuclear weapons should be used in limited wars in the Far East.[13]

A similar attitude was expressed by the Minister of Defence, Harold Macmillan in 1955:

> …the power of interdiction upon invading columns by nuclear weapons gives a new aspect altogether to strategy, both in the Middle East and in the Far East. It affords a breathing space, an interval, a short but perhaps vital opportunity for the assembly, during the battle for air supremacy, of larger conventional forces than can normally be stationed in those areas.[14]

These mid-1950s discussions pre-date Chinese acquisition of nuclear weapons a decade later, but formed a context for the later consideration of the deployment of nuclear forces in the Indian Ocean to provide a nuclear guarantee for India against China. The 1965 Defence White Paper did not specifically announce such a deployment but implied the need for reassurance of regional nonnuclear powers and this was widely assumed to involve regional nuclear deployments. Although it is nowhere specifically stated, the plain indication of the White Paper is that Britain intends to keep an independent nuclear force outside the NATO area, composed in the first place of V-bombers plus carrier-based aircraft. The main purpose of this force would be to provide a nuclear guarantee to India and other Commonwealth countries which fear Chinese (and, possibly in the future, Indonesian) nuclear

[13]Martin S. Navias, *Nuclear Weapons and British Strategic Planning, 1955–1958,* Clarendon Press, Oxford, 1991, p. 48.

[14]House of Commons, Volume 568, Column 2182, 2 March 1955.

blackmail and which might otherwise, in some cases, want to make their own nuclear bombs.[15]

Mention has already been made of the Royal Navy's nuclear-capable forces, which, in the 1980s, comprised WE-177 nuclear bombs for the Sea Harrier operating from any of three Invincible-class carriers, and the nuclear-depth bomb variant of the WE-177 carried on Anti-Submarine Warfare (ASW) helicopters. In this context, though, it is worth noting that naval nuclear deployments long pre-dated these weapons.

Prior to the deployment of nuclear-armed *Sea Harriers* from 1980 to 1992, there had been a two-year period when Britain had not had nuclear-capable fixed-wing aircraft regularly operating from aircraft carriers. Prior to 1978, however, there had been a 16-year period from 1962 in which Scimitar and Buccaneer aircraft had operated in a nuclear-capable role on carriers such as *Eagle, Centaur*, *Victorious*, *Hermes* and *Ark Royal.* There has been little open-literature discussion of the roles of these forces, either within the NATO context or elsewhere.

This description of the 'out-of-area' deployment of nuclear weapons by Britain is matched by various indications of a preparedness to consider using them in limited conflicts. As Milan Rai writes in his paper, Tactical Trident,[16] Sir John Slessor, Marshall of the RAF in the 1950s, and one of the most influential military theorists of the period, believed that 'In most of the possible theatres of limited war…it must be accepted that it is at least improbable that we would be able to meet a major communist offensive in one of these areas without resorting to tactical nuclear weapons' Official statements reflect a similar approach. The 1956 Defence White Paper remarked that while 'the consciences of civilised nations must naturally recoil from the prospect of using nuclear weapons… we have to be prepared for the outbreak of localised conflicts short of global war. In such limited wars the possible use of nuclear weapons cannot be excluded'.

Rai goes on to relate the circumstances of nuclear deployments in South East Asia in the early 1960s, arguing that this represented, in part, a low-level nuclear threat to Indonesia during the UK/Indonesian confrontation. He points to the deployment of strategic nuclear bombers to Singapore in December 1963 as an implicit threat to the Jakarta authorities. There appears to be some support for this in the views of the historian of the Far East Air Force, Air Chief Marshall Sir David Lee regarding the V-bomber deployment to Singapore, 'their potential was well-known to Indonesia and their presence did not go unnoticed'. He also commented, 'It is clear beyond doubt that the knowledge of RAF strength and competence created a wholesome respect among Indonesia's leaders, and the deterrent

[15]'Labour's Bomb and the White Man's Burden', *The Observer*, London, 28 February 1965.

[16]Milan Rai, 'Tactical Trident, the Rifkind Doctrine and the Third World', *Drava Papers*, London, 1994. Rai references his quotations fully.

effect of RAF air defence fighters, light bombers and V-bombers on detachment from Bomber Command was absolute'.[17]

Lee's assessment is, in turn, supported by a report from the Commander in Chief, Far East, Air Chief Marshall Sir John Grandy, concerning a later period of confrontation in 1965: 'four reinforcement medium bombers *(*Victor or Vulcan*)* were retained in the Command throughout the period. This ensured that all principal targets under contingency plans were covered with forces immediately available. There is little doubt that this force has provided a valuable deterrent to confrontation being conducted on a larger scale...'[18]

In a series of planned operations code-named *Addington*, V-bombers would participate in operations intended to eliminate Indonesian air strike capability. While Addington One assumed cooperation with Australian and New Zealand forces operating from RAAF Darwin, Addington Two assumed unilateral action.[19] Under these circumstances it is difficult to see how the Indonesian air strike capability could have been eliminated by four V-bombers armed only with conventional munitions.

It could, though, still be argued that Canberra and Victor deployments to Singapore represented a conventional force deterrent, but the dual-capable nature of the planes at least allowed the British to imply a greater potential, and this follows a pattern established by the United States as early as 1946.[20]

The willingness of British military planners to embrace the possibility of limited nuclear war in regional confrontations contrasts strongly with the early reluctance (at least until the 1960s) to accept the viability of the concept in Europe.[21] The risk of nuclear escalation in Europe was, of course, uncomfortably close to home, whereas distant regional conflicts were less directly risky to the UK itself.

In essence, though, during the mid and late 1950s, nuclear weapons outside Europe were seen as providing a cheap option for a country no longer able to support massive conventional forces. This was particularly true after the Suez crisis in 1956, and was expressed in the 1957 Defence White Paper of Duncan Sandys.[22]

The attitude of the times was perhaps best expressed by the Defence Minister, Duncan Sandys, in the 1957 Defence Debate:

'One must distinguish between major global war, involving a head-on clash between the great Powers, and minor conflicts which can be localised and which do not bring the great Powers into direct collision. Limited and localised acts of

---

[17]David Lee, Eastward *A History of the Royal Air Force in the Far Fast, 1945–1972,* IIMSO, London, 1984.

[18]Humphrey Wynn, *RAF Nuclear Deterrent Forces,* HMSO, London, p. 440.

[19]*Op. cit*., p. 444.

[20]Barry M. Blechman and Stephen S. Kaplan, *Force Without War – U.S. Armed Forces as a Political Instrument,* The Brookings Institution, Washington DC, 1978.

[21]*See Baylis, op. cit., pp. 331–337. See also, Ian Clark,* Nuclear Diplomacy and the Special Relationship, Britain's Deterrent and America, 1957–1962, *Clarendon Press, Oxford, 1994, pp. 120–125.*

[22]Michael Dockrill, *British Defence since 1945,* Blackwell, Oxford, 1988, pp. 65–81.

aggression, for example, by a satellite Communist State could, no doubt, be resisted with conventional arms, or, at worst, with tactical atomic weapons, the use of which could be confined to the battle area'.[23]

Almost exactly 25 years later, a localised act of aggression in the South Atlantic was to lead to the deployment of nuclear weapons in a potential war zone.

9.4 Nuclear Weapons and the Falklands War

Britain has a history of out-of-area deployment of nuclear weapons and has also had a long-term involvement in NATO nuclear planning for 'flexible response', including the early first use of nuclear weapons in a conventional conflict It is in this context that the deployment of nuclear weapons during the Falklands War in 1982 is particularly significant, not least for what it tells us about the possible future uses of Trident.

At the outbreak of the Falklands War, Britain's naval tactical nuclear weapons comprised free-fall bombs and depth bombs. The two operational aircraft carriers, *Invincible* and *Hermes*, carried both kinds of weapon, and over 40 destroyers and frigates could carry depth bombs for helicopter-delivery. At that time, it was acceptable for any of the warships to carry nuclear weapons in peace-time. The helicopter-delivered anti-submarine nuclear depth bombs were considered particularly significant weapons against fast Soviet submarines, and even the smaller Royal Navy frigates such as the Type-21 had crew fully trained in nuclear weapons operations.[24]

After the Falklands War, only the carriers and Type-22 frigates were certified for peacetime nuclear deployments, possibly as a result of experience in the war. The total number of naval tactical weapons in the inventory at the time was small, perhaps 25 nuclear depth-bombs and a similar number of gravity bombs, although the RAF also had over 100 of the latter.

When Argentina invaded the Falklands at the beginning of April 1992, a major political crisis faced the British government, which was already experiencing considerable political unpopularity as a result of domestic economic and social policies. The Royal Navy, too, was in a period of crisis, as the 1991 Defence White Paper had included substantial cuts in the surface navy, due to come into force progressively from 1992. Both government and navy therefore had much to lose, or gain, depending on the outcome of the crisis.

In the immediate aftermath of the Argentine invasion a very substantial naval task force was quickly assembled and some elements of it set sail from Britain within four days of the invasion. Others, including the destroyer *HMS Sheffield*, disengaged from a NATO exercise, Spring Train, in the western Mediterranean, and

[23]House of Commons, Volume 568, column 1765, 16 April 1957.

[24]Interview with the weapons officer of a Type 21 frigate, 1983.

headed for Ascension Island to link with the task force. The then Defence Minister, John Nott, specifically stated that warships were being deployed with their full range of weapons, which implied a nuclear capability. The precise phrase he used in the Falklands debate on 3 April, 1982, was that they were 'sailing under wartime orders and with wartime stocks of weapons'.

When defence ministers were subsequently questioned on whether nuclear weapons were being deployed, they fell back on traditional Whitehall phraseology which gave the impression that there was no risk of nuclear escalation without actually saying that there were no weapons deployed. Thus, Minister of State Lord Trenchard replied to a question on 19 April: 'There is no question but that nuclear weapons are not applicable to the current situation in the Falkland Islands area'.

Six days after the initial elements of the task force left Britain, *The Observer* reported that nuclear weapons were almost certainly embarked on some ships,[25] and that paper's defence correspondent, Andrew Wilson, later said that he talked personally to one frigate captain who was not prepared to leave his base and head for a war zone without nuclear weapons on board.[26] Other sources indicate that *HMS Sheffield* and other ships from Spring Train also went south carrying nuclear weapons. Wilson's write-up in *The Observer*, referred to the Task Force thus: 'It is almost certainly carrying tactical nuclear naval weapons – atomic depth charges carried by Sea King helicopters and free-fall bombs carried by Harrier jump jets – as part of its NATO equipment. Although it would be irresponsible at this stage to suggest that there is the remotest intention of using them against the Argentines, some defence experts (and not only Mr Tony Benn) are concerned at what might happen if the conflict were to escalate and bring in more powerful naval forces'.

Former Navy Minister, Keith Speed MP, who had resigned the previous year in protest at cuts in the navy budget later said that he, 'would have been astonished if those ships, from exercise Spring Train, had not been carrying nuclear weapons'.[27]

According to the Labour MP, Tam Dalyell, there was consternation in the Ministry of Defence when it was appreciated that a very large proportion of the Royal Navy's entire stock of nuclear weapons was heading for a potential war zone. While there might have been concern over possible escalation to nuclear use, there was certainly concern over the loss of such weapons in a conventional conflict. Senior staff at the Ministry of Defence were reportedly worried about losing any of the navy's relatively small stock of tactical nuclear weapons.

As Dalyell put it, 'There was a tremendous row about this inside government and Whitehall. As a result, some of the nuclear weapons were lifted back by helicopter and other boats before the Task Force reached the Western Approaches. The rest stayed on board'.[28]

---

[25]Andrew Wilson, 'Deadline Midnight', *The Observer*, London, 11 April 1982.

[26]Interview, 1984.

[27]Quoted in a paper by William M. Arkin and Andrew S. Burrows, *British Nuclear Weapons in the Falklands*, Published by the Institute of Policy Studies, Washington, DC, 1982.

[28]Tam Dalyell, 'Secrets of Davy Jones's Locker', *New Scientist*, 24 March 1983.

According to a member of the Royal Fleet Auxiliary, a number of the nuclear weapons which had got as far as Ascension Island on Task Force ships were off-loaded from warships on to the RFA *Regent.*[29] While this ship continued down to the South Atlantic carrying the tactical nuclear weapons on board, it was kept away from the main zone of conflict for most of the war, unlike its sister ship RFA *Resource*, which was forward-based, including a period in San Carlos Bay at the time of the amphibious landings.

As far as tactical nuclear weapons are concerned, it thus appears that a number were deployed on Task Force ships sailing from Britain but these were, at least in part, kept away from the main combat area, although still available for use in extremis. It is not known whether this deployment pattern also applied to nuclear weapons on the two carriers, *Invincible* and *Hermes.*

What then of *Sheffield* and the other ships from Spring Train which headed directly south? According to *The Times*, 'Some British ships in the South Atlantic during the Falklands campaign were carrying nuclear anti-submarine weapons. Whitehall sources said that some of the frigates which went to the Falklands had been involved in exercises in the Mediterranean, and would have been routinely carrying anti-submarine nuclear bombs. Because they were diverted directly to the South Atlantic there would have been no opportunity to offload the weapons'.[30]

There were many reports, but really amounting to no more than rumours, that *Sheffield* had WE-177 nuclear depth bombs on board when hit by the Exocet missile on 4 May, 1982. Some reports surfaced in the press, and others spoke of *HMS Coventry*, sunk on 25 May, also having nuclear weapons on board. There was further speculation that the extensive salvage operations conducted on the wrecks of *Coventry* and *Sheffield* were concerned with recovering these weapons.

With regard to *Coventry*, it is more likely because it sank so quickly after being bombed, a matter of less than 30 min, that it was not possible to destroy some advanced equipment, possibly including code systems, and it was these which were retrieved from the sunken ship, which had not been carrying nuclear weapons. The case of *Sheffield*, though, may be different. There is some evidence to indicate that it was carrying nuclear depth bombs[31] but this is largely circumstantial and has been strongly denied in some circles. One other aspect of British deployments in the Falklands is relevant. One of the earliest actions against Argentine forces was the bombing of the runway at Port Stanley airfield by a single *Vulcan* bomber operating from Ascension Island, nearly 4,000 miles away. This very long-range and arduous operation involved multiple air-to-air refueling of an aircraft carrying conventional HE bombs. The purpose was to disrupt the runway, but a raid by one plane was highly unlikely to cause sustained damage, as indeed proved to be the case. The

---

[29]Interview, 1985.

[30]*The Times*, 3 November 1982.

[31]For a more detailed account of the *HMS Sheffield* question, see Paul Rogers, 'A Note on the British Deployment of Nuclear Weapons in Crisis - with particular reference to the Falklands and Gulf Wars and the Purpose of Trident' *Lobster*, Number 28, 1994, pp. 2–10.

Vulcan raid, though, did demonstrate that Britain had the capability to deploy nuclear-capable aircraft over a range which brought almost the whole of Argentina within reach.

Using the more reliable sources, it would appear that tactical nuclear weapons were deployed on Task Force ships leaving Britain and also on some of those diverted from exercise Spring Train. Most, but probably not all, were removed from deployment in the war zone, but primarily because there were fears that too many of the navy's small stock would be lost. Even those kept back from the immediate war zone remained readily available during the conflict.

It should be added, though, that the utility of these particular nuclear weapons was somewhat limited, in a purely military sense, in the context of the Falklands War. Argentina had two small, quiet German-built Type 209 submarines which caused the Royal Navy some considerable concern, but the latter's tactical nuclear depth bombs were designed to be used against large and thick-hulled Soviet submarines operating in deep open water, not against small submarines in fairly close proximity to Royal Navy warships, though at least one naval officer still felt they were sufficiently useful to insist on taking them. Similarly, the *Sea Harriers* on the aircraft carriers may have been capable of delivering nuclear gravity bombs, but targets on the Falklands were out of the question and targeting anything on the Argentine mainland would involve difficult problems of range and survival for the aircraft. Against that, Argentina did have two large surface warships, the aircraft carrier *25th May*, and the heavy cruiser, *General Belgrano.* The former took little part in the war and the latter was sunk by conventional torpedoes, but these aspects were not known to planners prior to the actual war.

It should be remembered, though, that the decision to deploy a major task force to the South Atlantic was a singularly high-risk strategy, certainly the most risky large-scale UK military operation since Suez in 1956. Its essential purpose was to facilitate the landing of a substantial force to re-take the Falkland Islands, a purpose which would have been rendered virtually impossible by the loss of either of the two aircraft carriers or the troopship *Canberra.* The military and hence the political stakes were thus high.

In relation to tactical nuclear weapons, the available evidence does appear to show that elements within the navy were readily prepared to see such weapons deployed directly in a war zone but that, to some extent, wiser counsel prevailed and at least some were kept at a distance. This was not, however, the case with Britain's other naval nuclear system, the strategic Polaris missile.

At the time of the Falklands War, Britain had four Polaris submarines, sufficient to keep one on patrol in the North Atlantic at all times, with a second available for possible additional deployment. The regular patrol was aided by one or two nuclear-powered hunter-killer submarines for protection, a policy known as deterrence support. The Polaris submarines themselves carried torpedoes but these were essentially for self-defence.

In the two years after the war, information from several highly-placed sources indicated that a Polaris missile submarine had been diverted to a patrol area in the mid-Atlantic several thousand miles north of the Falklands but nevertheless one which allowed it to operate within missile range of Argentina.

Because of the significance of this deployment, it is worth saying a little about the nature of these reports as the story has been subject to some ridicule by some sectors of the defence establishment. Tam Dalyell's initial source, shortly after the Falklands War, was a senior Conservative MP with an interest in defence matters and close links with the Ministry of Defence. Dalyell later had it confirmed to him by a senior officer in the Polaris fleet who retired after the Falklands War.[32] A retired senior Ministry of Defence civil servant was a further source of information.[33]

Two years after the war, in 1984, journalists working for the *New Statesman* were informed of classified signals exchanged between London and the British Embassy in Washington concerning the deployment. Duncan Campbell and John Rentoul published this information as part of the 'Belgrano Papers' issue of the magazine shortly after Clive Ponting was arrested for a quite different matter – leaking details of the attempt by the government to mislead a House of Commons Select Committee over the *Belgrano* sinking. As they put it:

> One well-placed political source has already revealed to Tam Dalyell that a Polaris submarine was sent to the South Atlantic. Dalyell was informed that the submarine went as far south as Ascension; the likely target for a threatened or demonstration nuclear attack was said to be Cordoba in northern Argentina. The nuclear threat might have been used if any of the task force's capital ships - one of the carriers or the troop ship *Canberra* - had been destroyed in a missile attack. The Polaris deployment was said to have been ordered in the wake of the sinking of *HMS Sheffield*, after ministers had to confront the possibility that Argentine air superiority and Exocet missiles could mean the military defeat of the British task force and the rapid political extinction of the Thatcher government.[34]

The New Statesman has been able to confirm that a Polaris submarine was indeed deployed to this position. Details of the deployment are given in a series of highly classified telegrams sent to the British Embassy in Washington.'

This report of the Polaris deployment formed one part of a much longer article which was primarily concerned with the *Belgrano* affair and included publication of a classified document by the magazine. It was these aspects of the story which captured most media attention, although the Polaris deployment did figure fairly prominently.

One interesting indication of its veracity came in the response of MoD officials to press enquiries immediately after the magazine's publication According to MoD sources, there might well have been a Polaris submarine deployed to the South Atlantic, but this was because the navy was short of hunter-killer submarines and a Polaris missile submarine can serve this function.[35]

---

[32]Interviews with Tam Dalyell, 1982–4.

[33]Interview, 1985.

[34]Duncan Campbell and John Rentoul, 'All Out War', *New Statesman*, 24 August 1984.

[35]Interview with BBC journalist, 1984.

Such an explanation is little short of fanciful – while Polaris submarines do have a limited hunter-killer capability, this is for self-defence and they are normally escorted by dedicated hunter-killers. To risk one of Britain's four strategic missile submarines in a war zone in this matter would have been hugely risky. Since Britain needed four Polaris submarines to maintain at least one on patrol at all times, risking a submarine would have meant risking the entire Polaris capability.

There is another piece of supporting evidence which also relates to hunter-killer submarines. Reports from ex-servicemen involved in the Falklands War have made it clear that there was a severe shortage of hunter-killer submarines to form a protective screen around the Task Force. Whereas the MoD gave an impression that five such submarines were available, for much of the Falklands War only two, or at most three, were actually deployed. There are indications from MoD sources that the shortage was due to the use of two hunter-killer submarines to act as protective escort to the Polaris missile submarine deployed in mid-Atlantic to cover Argentina.

Official confirmation of the Polaris deployment will probably have to wait at least for the 30-year rule, but the copious leaks are reasonably conclusive. Campbell and Rentoul are probably right in their comment that this extreme measure – deployment of a substantial nuclear system against a non-nuclear power, was in the context of possible defeat of the Task Force and consequent collapse of the Thatcher government. Extreme circumstances appeared to justify extreme measures, but the whole affair also has to be seen in the context of the British nuclear establishment's long term attitude to nuclear weapons. Contrary to popular opinion, nuclear weapons are seen as useable instruments of war, and there is a long-standing belief that they are appropriate for use in crises involving nuclear or non-nuclear states. It is a theme which emerged again in the conflict in the Gulf, nearly a decade after the Falklands War. It is also singularly relevant in relation to the roles which the Trident system may be expected to play in future regional crises.

9.5 The Gulf War

Following the Iraqi invasion of Kuwait in August 1990, and in the run-up to the war itself, which commenced on 16 January 1991, there was considerable concern over possible use of chemical and possibly biological weapons by Iraq against coalition forces. A number of western political sources hinted at a nuclear response to any substantial Iraqi CW use, and it was widely assumed that the routine equipping of US warships, especially the nuclear-powered aircraft carriers, at least gave the US a nuclear-weapons capability in the immediate vicinity of Iraq.

In Britain, in the weeks leading up to the war itself, there were a number of indications of a willingness to escalate to nuclear use, although most of these came from non-official sources. Even so, there were some quite clear hints from government sources, junior defence ministers Archie Hamilton speaking of massive retaliation against Iraq if chemical weapons were used.[36]

The clearest indication of British willingness to use nuclear weapons came in a report in *The Observer* on 30 September, 1990, quoting a senior army officer attached to the 7th Armoured Brigade which had begun to leave for the Gulf the previous day. He confirmed that an Iraqi chemical attack on British forces would be met with a tactical nuclear response.

The deployment of nuclear weapons to the region by Britain presented difficulties. US nuclear weapons were deployed at sea, or at existing bases in the region such as Incirlik in Turkey. Britain had strike aircraft in Bahrain and Oman, but basing nuclear weapons in these states might not have been acceptable, and this could also apply to RAF Akrotiri in Cyprus.

The alternative was on board ships, but the only Sea Harrier-capable aircraft carrier despatched to the area was *HMS Ark Royal,* and this ship did not pass through the Suez canal during the war, being restricted to the Eastern Mediterranean. Its *Sea Harriers* were not, therefore easily within range of Iraqi territory should Britain have decided to undertake or take part in nuclear operations against Iraq.

It appears, though, that a rather remarkable alternative option was planned and may well have been implemented. In addition to the three Sea Harrier-capable aircraft carriers in the Royal Navy such as *Ark Royal,* there is just one other ship which can deploy these planes, the Royal Fleet Auxiliary *Argus.* RFA *Argus* is a large 28,000-tonne auxiliary vessel which is officially classified as an aviation-support ship, and converted in the mid-1980s from the container ship *Contender Bezant.* It was originally used as an aircraft delivery ship during and after the Falklands War and was purchased by the MoD for conversion at Harland and Wolff, Belfast, in 1984, entering service in 1987. It can carry up to 12 Sea Harrier aircraft and also Sea King helicopters. It will be replaced in the late 1990s by the purpose-built helicopter carrier *HMS Ocean.*

Argus was deployed to the Gulf during the crisis, officially as an auxiliary ship in the role of forward casualty evacuation point. The impression was actually given to the media that *Argus* was a hospital ship, but it was never designated as such (e.g. red cross and white paint) and was militarily operational during the war. Although its prime use was for receiving British casualties, of which there were fortunately very few, it also carried a wide range of stores.

Prior to the actual war, detailed planning was certainly done to make it possible for the ship to be equipped with tactical nuclear weapons,[37] although they may even have been actually deployed to the Gulf. If so, it would have represented a direct British nuclear armed presence in the war zone.

How the weapons would have been deployed, had the decision to do so been taken, is not clear. One option would have been to use Sea Harriers, another would

[36]Reported, for example, in *The Observer,* 6 January 1991.

[37]Interviews, 1993 and 1994.

have been to air-lift the nuclear bombs to a Tornado-operating base on the mainland. It is assumed that nuclear use would have been restricted to a response to large scale Iraqi CW attacks on coalition forces requiring tactical nuclear use against Iraqi artillery and missile concentrations. US or British free-fall nuclear bombs on strike aircraft such as the F-15, F-18, Tornado or Sea Harrier would have been considered the appropriate means of delivery. Even without *Argus,* Britain would have had a small nuclear capability in the Mediterranean, and could have moved nuclear bombs for Tornado strike aircraft from Britain or Germany short notice.

In the aftermath of the Gulf War, suggestions that either the United States or Britain had any intention of deploying or using nuclear weapons in the conflict were largely ridiculed by politicians and senior military officers, who were at pains to point out that there had been little risk of the Iraqis using chemical or biological weapons. This contrasted with the extensive efforts made to protect coalition troops against the effects of chemical and biological weapons.

Four years later, new revelations about the war gave a rather different picture. The US military commander at the time of the war, General Colin Powell, published his autobiography in 1995, confirming that the then Defense Secretary, Dick Cheney, had required him to draw up plans for the use of nuclear weapons against Iraq.[38] Even more remarkable was the evidence provided in August 1995 to the UN Commission investigating the status of weapons of mass destruction in Iraq. This showed that, prior to the war, Iraq had produced large quantities of biological as well as chemical weapons, and these included anthrax bombs and warheads for Scud missiles.[39]

Considerable further evidence was published by the UN Special Commission (UNSCOM) in October and December 1995, and this is worth some examination, partly in relation to the Gulf War itself, but also because of its longer term implications. The October 1995 report[40] reveals a remarkably comprehensive biological weapons (BW) programme, far in advance of anything previously revealed in public, along with more advanced chemical weapon (CW) nuclear and missile programmes. The December 1995 report[41] describes Iraq's development and testing of radiological weapons.

The Iraqi BW programme commenced as early as 1975 and received considerable technical and financial investment in the mid and late 1980s. Research was undertaken on numerous causal organisms of human and crop diseases, with emphasis placed on anthrax, botulinum toxin and aflatoxin. Over 11,000 litres of botulinum toxin and 8,500 litres of anthrax spore suspension were produced by early 1991, extensive field trials of several weapons systems were undertaken, and

---

[38]Reported in *The Washington Times,* 'Cheney discussed nuke use in the Gulf, 10 September 1995.

[39]'Iraq's Secret Arms Programmes Bigger than Once Thought', Agence France-Presse International News, Dubai, UAE, 23 August 1995. See also, Jack Redden, 'Iraq had Germ War Missiles Before Gulf War', *Reuter News Reports,* Amman, Jordan, 23 August 1995.

[40]Report to the Security Council, S/1995/865, 11 October 1995.

[41]Report to the Security Council, S/1995/1038, 17 December 1995.

anthrax, botulinum and aflatoxin were all weaponised by January 1991. The Iraqi CW programme included work on the advanced V-agents.

At the outbreak of the Gulf War, the Iraqis had produced 166 R400 bombs filled with botulinum, anthrax or aflatoxin, and a special production run of the A1 Husayn ('Super-Scud') missile in 1990 had resulted in 25 missiles with warheads filled with the same BW agents.

More significantly, in early January, 1991, immediately prior to the outbreak of the war, the biological weapons, both bombs and missiles, were dispersed to four locations, along with a number of missiles loaded with chemical warheads. Authority to launch these weapons was pre-delegated to regional commanders in the event that Baghdad was hit by nuclear weapons. Pre-delegation of launch authority for weapons of mass destruction in time of war is potentially unstable and highly dangerous. It is certainly possible that a coalition move against Baghdad in the closing stages of the war might have elicited a CBW response from a regional commander. Given that BW systems, in particular, are capable of causing casualties broadly similar to a nuclear attack, a coalition nuclear response would have been possible.

It is appropriate to note that a major war game exercise at the US Naval War College in July 1995 followed the scenario of a two-fold crisis involving North Korea and a re-emergent Iraq. Both crises escalated to CW use by these states. In the war game, the North Korean crisis was eventually contained, but the Gulf crisis escalated to a major Iraqi BW attack on Dhahran with a retaliatory US nuclear strike against Baghdad.[42]

Although information on the Iraqi BW programme only became available to the public with the publication of the UNSCOM reports, most of it was known to the US Department of Defense in 1990. An Inter-Agency Intelligence Community working group on Iraqi CBW reported to the Defense Secretary in December 1990 on the Iraqi BW programme.[43] In addition to confirming the weaponization of botulinum and anthrax, it reported that BW agents were dispersed to 35 bunkers at 17 locations throughout Iraq, and that 'Iraq would consider using BW as a weapon to save the regime from falling'.

Although it is now five years since the Gulf War, revelations about the Iraqi CBW programme are likely to cause some major reassessments of the conflict and its implications. Iraq was able to develop significant weapons of mass destruction, and it must be assumed that other states are also able to do so. This is already having an effect on considerations of the wider nuclear context in a number of nuclear weapon states including Britain.

---

[42]Theresa Hitchens, 'Wargame Finds U.S. Falls Short in Bio War', *Defense News*, Washington DC, 28 August 1995.

[43]Reported in *The Chemical Weapons Convention Bulletin*, Number 30, December 1995, Science Policy Research Unit, University of Sussex.

9.6 The Wider Nuclear Context

The role of the British Trident system is developing within the context of wider nuclear developments principally involving the United States, but also relating to developments in Russia and France. It is appropriate to discuss British developments in terms of this more global context.

In the United States, the nuclear debate has evolved through three reasonably clear phases since 1990. The first phase, lasting until early 1993, was essentially a process of looking for new tasks for existing nuclear forces, usually a combination of fears of a resurgent Russia and of perceived third world threats against US interests This was reflected, early in 1991, in the publication of a draft of the Strategic Deterrence Study undertaken for US Strategic Air Command (also known as the Reed Report). Its terms of reference stated the belief that 'the growing wealth of petro-nations and newly hegemonic powers is available to bullies and crazies, if they gain control, to wreak havoc on world tranquillity'.[44] The study itself called for a new nuclear targeting strategy which would include the ability to assemble 'a Nuclear Expeditionary Force, primarily for use against China or Third World targets', which would be required because 'Nations with the wealth and ideological fervour to pursue nuclear programs, no matter what the time or cost, are very different' from traditional nuclear powers such as Britain and France. North Korea, Algeria, Libya, Iran and of course Iraq fit this bill. To quote: 'They and their terrorist cousins are more likely driven by…the desire to…terrorise, blackmail, coerce, or destroy' among other motives.[45]

Two years later, the *New York Times* reported that staff' at US Strategic Command (the successor to Strategic Air Command which includes ballistic missile submarines) were 'in the early stages of building and testing computer models that could enable Mr Clinton to aim nuclear weapons at third world nations that threaten the interests of the United States and its allies.'[46]

Along with these developments in strategy, there were indications of new nuclear weapon research and development programmes aimed at producing weapons appropriate to new tasks. They included small nuclear warheads suitable, for example, for destroying missiles armed with biological warheads, others for attacking deep underground bunkers, as well as electro-magnetic pulse (EMP) warheads for disabling electronic equipment.

By mid-1993, though, a second phase was developing, in part in reaction to these evolving nuclear ambitions. The Clinton administration, formed at the end of 1992, included a number of members of the Washington arms control community,

[44]Strategic Advisory Group of the Joint Strategic Target Planning Group, US Strategic Air Command, 'The Role of Nuclear Weapons in the New World Order'. Reported in *Navy News and Undersea Technology*, Washington DC, 13 January 1992.

[45]Ibid.

[46]Eric Schmitt, 'Head of Nuclear Forces Plans for a New World', *New York Times*, p. B7, 25 February 1993.

giving the community itself more influence than it had at any time since the Carter administration more than a decade earlier. This, coupled with a relatively liberal Congress and the work of a number of academic analysts, gave a new emphasis to ideas of a nuclear-free world, ensuring a debate which moved the centre of gravity away from the more hawkish nuclear view.[47] It is perhaps relevant to note the almost total absence of an equivalent debate in Britain, at least in academic circles.

This changed climate of opinion in the US was boosted by the relative openness of the Department of Energy under Hazel O'Leary'. In these circumstances, opinion in Congress was sufficiently strong to curb a number of budget proposals for research and development of new nuclear systems. By late 1994, the administration could claim that these programmes were on ice, although some analysts suspected that work continued to be funded in classified or 'black' programmes. While the Nuclear Posture Review of September 1994 made clear the administration's continuing commitment to nuclear weapons, its presentation and outlook was certainly far less hawkish than the outlook at the end of the Bush administration two years previously.

Already, though, the debate was entering a third phase of renewed support for the enhancement of nuclear forces to meet new perceived threats, a process helped in no small measure by the Republican control of Congress after the November 1994 mid-term elections. The possibility of a resurgent Russia was cited as the main reason for maintaining effective nuclear forces, allied to a perceived need to counter Chinese capabilities. In the latter part of 1995, though, the most important single motivation became the risk of biological weapons being developed by supposedly 'rogue' states, Iraq, Iran, Libya and North Korea being the most oft-cited examples.[48]

In September 1995, there were reports that the United States was[49] commencing a two-year programme to modify one of its standard strategic nuclear bombs to enable it to take on an earth-penetrating role, able to destroy command bunkers or CBW facilities which were too well protected to be destroyed by conventional weapons. While not technically a new weapon, this enhancement of an existing system was being undertaken very much in relation to perceived threats from a proliferating world.

This transition from strategic nuclear targeting in the Cold War to selective nuclear targeting in the future is considerable. The Cold War use of strategic nuclear forces would involve a large-scale launch of weapons against a wide range of targets or aim-points, most of them known and analysed months or years in advance and all forming part of the Single Integrated Operational Plan (SIOP).[50]

---

[47]For example, Barry M. Blechman and Cathleen S. Fisher, 'Phase Out the Bomb', *Foreign Policy*, No. 97, Winter 1994–95, pp. 79–95.

[48]See reference 41.

[49]'Sandia Redesigning N'arsenal Weapon', *Albuquerque Journal*, p. Al, 12 September, 1995.

[50]James H. Patton JR., 'New Roles on Horizon for Triad's Last Leg?', *International Defense Review*, September 1994, pp. 38–42.

As the ex-Soviet target list declines, the number and distribution of aim-points will be much less constant and may occur anywhere in the world, defying pre-planned targeting. A recent analysis of third world targeting, and of demands likely to be made by the US National Command Authority (NCA), points out that, 'Some of these targets may warrant, under specific but undefined circumstances, a rapid response with a single low-yield warhead. As the SIOP as we have known it becomes smaller, the recently established US Strategic Command (STRATCOM) may find itself required to generate a similar attack plan quickly – using its precise database of all 'potentially strategic targets in the world' – matched to a specific scenario on demand from the NCA or a commander-in-chief (CINC). Such a plan might consist of only a few targets, to which the NCA would match the appropriate weapons and delivery systems'.[51] The United States may disarm and ultimately dismantle most of its Cold War arsenals of nuclear weapons, yet, on present trends, will retain nuclear forces still numbered in the thousands, adapting its targeting and strategy to an uncertain world.

These motives, especially a concern over CBW proliferation, apply also to current Russian strategic thinking, but they are joined by a feature particular to the Russian context. With the rapid decline of the Russian economy, it has proved quite impossible to maintain Russian conventional forces at anything remotely like their Cold War levels. Moreover, the conventional forces which have been retained are seriously deficient in capability, a predicament shown clearly in the great difficulty encountered in controlling the Chechnya rebellion in 1995. In the absence of effective conventional forces, nuclear weapons are increasingly being seen to provide a much cheaper military option in the event of serious security threats to Russia.[52]

More generally, nuclear forces are seen as almost the only way of maintaining any semblance of military superpower status, especially in the face of NATO influence moving steadily eastward These trends, in combination, help explain a continuing Russian commitment to maintaining core strategic nuclear forces, including modernising some systems, and preparing to deploy tactical nuclear weapons close to the western borders of Russia.

All of these tactics are, to an extent, bound up with the complex machinations of Russian politics, and they appeal most strongly to the weakened military and the more extreme nationalists. They are not restricted to such groups, however, and the nuclear option is more widely seen as one of the few sources of status, and indeed concrete military power, left to an economically crippled if formerly strong state. This is happening, moreover, at a time when the overall nuclear weapons infrastructure remains in disarray, with inadequate security of stored weapons and a potential leakage of expertise as unpaid or poorly paid nuclear weapons specialists

[51]Ibid.

[52]Theresa Hitchens and Anton Zhigulsky, 'Hard-Line Russians Tout Nukes to Match West', *Defense News,* 20 November 1995, p. 1. See also: 'Russia Builds up Nuclear Arsenal as Prospects for START II Fade', *Defense News,* Washington DC, 4 December 1995, p. 40.

seek to maintain their personal security. Outside indications suggest that within Russian military circles, a rigorous process of selective maintenance of nuclear systems is under way, ensuring that a core nuclear force involving strategic and tactical weapons is available, even if most of the nuclear weapons infrastructure is in disarray.

The developments in the United States, with an emerging debate about post-Cold War nuclear strategy, are not dissimilar to those in France.[53] This involves two reasonably clear-cut views of the future directions in policy and weapons developments. The broadly centrist view leans towards minimal deterrence, an avoidance of limited nuclear options and a concentration on policies to promote non-proliferation. It contrasts with a more Gaullist view which emphasises the defence of French and allied interests through robust and flexible military capabilities – both offensive and defensive. This includes the development of nuclear forces capable of being used, if necessary, with control and discrimination, particularly in confrontation with countries of 'the South' that may be armed with nuclear, chemical or biological weapons.[54]

Chirac's victory in the 1995 presidential election resulted in the Gaullist view prevailing, and a high-profile and controversial decision to resume nuclear testing in September 1995. The test programme was designed partly to complete the development of two warheads and partly to ensure that French technicians could simulate tests of new warheads, should a comprehensive test ban treaty be negotiated.

In 1992, Britain and France created the joint commission on nuclear policy and doctrine to try to harmonise their policies in the post-Cold War world environment. In November, 1993, the then British Defence Secretary, Malcolm Rifkind, reported that 'there are no differences between France and the UK on the fundamental nuclear, issues.[55]

While there may be no fundamental differences on policy, France plans a more diverse nuclear force than Britain. In addition to its ballistic missile submarines, France is likely to maintain land and sea-based nuclear capable strike aircraft, currently the Mirage IVP and Super Etendard respectively, but ultimately two versions of the Rafale equipped with nuclear-armed land-attack missiles. French nuclear strategy is particularly geared towards counter proliferation. According to a recent report: French officials said the Rafale armed with a long-range standoff missile would be well-suited to deter the growing number of countries in the developing world that have acquired, or are acquiring, nuclear, bacteriological and chemical weapons as well as ballistic missiles.[56]

---

[53]Discussed fully in David S. Yost, 'Nuclear Debates in France', *Survival,* Vol. 36, No. 4, Winter 1994–95, pp. 113–119.

[54]*Op. cit.*, p. 114.

[55]UK Defence Strategy. A Continuing Role for Nuclear Weapoas', speech by Malcolm Rifkind at the Centre for Defence Studies, King's College, University of London, 16 November 1993.

[56]Giovanni de Briganti, 'France to Replace Mirage IVP Bomber Fleet by 1997', *Defense News,* Washington DC, 8 January 1996.

This 'suitability' of such a combination stems from having a low-yield accurate nuclear system appropriate for precision attacks on military installations, a move away from the more traditional French concentration on counter-city targeting.[57] The French approach to post-Cold War nuclear strategy therefore follows closely those of the United States and Britain in this respect.

Two further aspects of French nuclear policy are relevant. One is the common view that France needs its own nuclear forces because it can in no way rely on the United States for a nuclear umbrella in a disorderly world. The second is that its conventional forces are such that it will rarely be able to respond to a major regional crisis on its own unless it can supplement its conventional forces with a nuclear capability.[58] This latter argument echoes British thinking dating back to the 1950s.

9.7 In Relation to Trident

In both the Falklands and Gulf Wars, Britain had the means to escalate to nuclear use, as it apparently had had during the much earlier Indonesian confrontation. This should not come as any great surprise, since it forms part of a continuum in military thinking about nuclear weapons which certainly has parallels in the US, the former Soviet Union, and France, as well as being clearly represented in NATO's planning for early first use of nuclear weapons.

This continuum has, in a sense, three phases. The first was in the 1950s, especially after the post-Suez Duncan Sandys defence review which placed such heavy reliance on nuclear forces. At this time, there was a fairly open discussion of the use of nuclear weapons in conflicts short of all-out East-West war. By the end of the 1950s, though, the rise of the anti-nuclear movement in Britain made it rather less appropriate to draw attention to such strategy and tactics, and nuclear war-fighting attitudes 'went underground' somewhat for a couple of decades. They surfaced prominently in the 1980s as more information came out on NATO nuclear planning and its emphasis on early first use of nuclear weapons.

With the ending of the Cold War, and the diminishing of any risk of massive Soviet nuclear retaliation to limited nuclear war-fighting, the utility of nuclear use against third world states is coming to the fore, with a particular emphasis on submarine-launched nuclear weapons such as Trident. Britain is in a somewhat different position to the United States, Russia and France in that it is withdrawing all its nuclear forces except Trident by the end of 1998, and this is what makes Trident so interesting in the context of a 'sub-strategic' or tactical role. The Trident system – four submarines, each of which can be armed with 16 missiles carrying up

[57]Karin von Stokirch 'Revamping the Rationale for French Nuclear Deterrence', Newsletter of the Strategic and Defence Studies Centre, Australian National University, Canberra, Spring 1995, pp. 2.

[58]David S. Yost, 'France's Nuclear Dilemmas', *Foreign Affairs,* Volume 75, Number 1, pp. 108–118, 996.

to 6 warheads each – is regarded as a powerful multi-purpose system which may have been designed for the Cold War years but is eminently adaptable to be the 'all-singing, all-dancing' missile for a disorderly world.

In this new context, there are considered to be four levels of possible use of Trident.[59] The most substantial is a response to a use of nuclear forces against British forces, most likely in a regional crisis in the Middle East. A re-run of an Iraq/Kuwait-type crisis is an example. Secondly, Trident might be used in response to chemical or biological weapons. A third possibility would be a demonstration shot to deter an enemy from taking a particular course of action. Finally, a nuclear response might be made to a state which had refused to desist from a specific course of action. Three out of these four cases would involve nuclear first-use.

The transition from a Cold War role for Trident was described succinctly in *International Defence Review* where it is pointed out that the availability of Trident: means that the UK will obtain a global, nuclear, sub-strategic strike capability at remarkably low cost. It is also a system that is inherently more flexible and much less vulnerable than using aircraft. Thus, the task of the UK's strategic planners in 1994 is to develop a minimal force that will have strategic – and now sub-strategic – validity in the highly uncertain world through to 2024, and possibly well beyond.

It should be made clear that there continues to be a difference between Britain's declaratory policy on nuclear use and the actual deployment policy. Just as in the Cold War era, the official line is that nuclear weapons are weapons of last resort deterrence, and that their use in a limited nuclear war is virtually inconceivable In the Cold War years, this went hand in hand with a deployment policy which was linked both to NATO's early first use strategy and to the deployment of nuclear forces in regional crises such as the Falklands War. In the post-Cold War years, the government points to the withdrawal of Britain's tactical nuclear warheads as an indication that we are moving to minimal deterrence, yet NATO reports; 'The United Kingdom briefed us on the implementation of its decision to utilize the flexibility of its submarine-launched Trident ballistic missiles to undertake sub-strategic as well as strategic roles and thereby progressively to replace the capability now provided by its air-delivered nuclear weapons'.[60]

On occasions, though, it is appropriate to nuance the declaratory policy, a memorable example being the presentation of the case for the legality of nuclear weapons at a hearing of the International Court of Justice at the Hague in December, 1995. The UK Attorney-General, Sir Nicholas Lyell argued that nuclear weapons were not inherently indiscriminate;

> 'Much of the writing on nuclear weapons on which these arguments rely dates from the Fifties and early Sixties. Modern nuclear weapons are capable of far more precise targeting and can therefore be directed against specific military targets…' In some cases, such as the use of low-yield nuclear weapons against warships on the high seas or troops in sparsely

---

[59] Ibid.

[60] Press Communique M-DPC/NPG-1(95)57, Paragraph 23, 8 June 1995, Nato Press Service, Brussels.

> populated areas, it is possible to envisage a nuclear attack which caused comparatively few civilian casualties. It is by no means the case that every use of nuclear weapons against a military objective would inevitably cause very great collateral civilian casualties.[61]

Three things are apparent. Firstly, Britain is intending to deploy Trident in a multipurpose role including a capability to use it in a limited nuclear war. Secondly, there is a long history of Britain preparing for limited nuclear war, both independently and within NATO, dating back 40 years. Finally, Britain has a habit of deploying nuclear weapons in regional crises involving non-nuclear powers.

In seeing a continuing utility for nuclear weapons, Britain is adopting policies similar to the United States, China, France and Russia, not to mention states such as Israel and Iraq. While there is a public commitment to a comprehensive test ban and extension of the non-proliferation treaty, this does not extend to any commitment whatsoever to comprehensive nuclear disarmament. Quite the reverse – there is a firm belief in the need for nuclear forces and their utility in regional crises. It is because this view is shared with France that the United Kingdom has been quietly supportive of French nuclear testing.

Of the many problems raised by this outlook, probably the most important is the message it sends to would-be possessors of weapons of mass destruction. Britain and the other major powers remain firmly committed to these systems yet have a prime policy of attempting to deny other states similar capabilities. It is fundamentally self-defeating and is likely to be the biggest single obstacle to controlling weapons of mass destruction.

Furthermore, the idea of being able to respond to the proliferation of weapons of mass destruction by constructing a stable system of multi-directional deterrence is itself dubious – the flawed application of a simplistic bipolar model of deterrence to a far more complicated and uncertain world. The Iraqi experience shows that it is possible for determined states of intermediate technological ability to develop weapons of mass destruction, especially biological weapons. This suggests a considerable capacity for weapons proliferation in a disordered world in which a few powerful states retain the nuclear option while attempting to deny it to others.

If the circumstances were to arise in which the interests of a state such as Britain, France or for that matter the United States or Russia, were so threatened as to elicit a nuclear response against another state, there is every likelihood that this crossing of the nuclear threshold would elicit some form of retaliation. Such retaliation might not be immediate and would be unlikely to be of a direct military nature. More likely would be a radical indirect response in the form of attacks against cities using, for example, biological weapons.

This is not how the matter tends to be seen in western military circles. To return to the US Naval War College war-game cited earlier which ended in a US nuclear strike on Baghdad in response to a massive Iraqi BW attack; in commenting on the outcome, *Defense News* reported: 'The United States has virtually no response to the use of such potentially devastating weapons other than threatening to use

[61]N-Bomb Need Not Kill Many, says Minister', *The Observer*, London, 17 December 1995.

nuclear weapons, a Joint Staff officer said. But it is unclear whether even nuclear weapons could provide a deterrent, unless the U.S. was willing to take the difficult moral step of destroying a city, he said. On the other hand, if the United States did launch a nuclear attack in response, 'no country would use those weapons for the next 100 years' the official said'.[62] Quite the contrary, the crossing of the nuclear threshold would be much more likely to result in covert biological warfare attacks, not on Dhahran, but New York, Chicago or Washington.

In essence, the belief that a proliferated world can be controlled with weapons of mass destruction is a dangerous illusion. The processes and causes of proliferation have to be addressed. This raises questions of our understanding of international peace and security which are likely to transcend traditional strategic studies thinking. Indeed, it can be argued with some force that the response within the strategic studies community to a disorderly world is classically neo-realist, with the primary emphasis on enhancing security by military means.

There are, though, important exceptions to this prevailing outlook – the debate on the prospects for a nuclear-free world is far from over. The closing months of 1995 saw the establishment by the Australian Government of a commission to study prospects for a nuclear-free world, a decision largely consequent on public reaction to the French nuclear tests in the Pacific. In Washington, meanwhile, the continuing work of the Stimson Center's Project on Eliminating Weapons of Mass Destruction resulted in a report advocating an evolutionary US nuclear posture progressing through minimal deterrence to the eventual elimination of all nuclear weapons within a multilateral framework. The steering committee for the Stimson project is chaired by a former Supreme Commander, Allied Forces in Europe (SACEUR) and includes three other retired generals.[63]

In the coming years there is therefore the prospect of an intense debate on the future role of weapons of mass destruction weapons, a debate which should best be considered in the wider context of issues of North-South relations, centering on a concern with developmental and environmental security. This is a considerable task but it is a debate which is beginning to take shape, at least in the United States. What is unfortunate is its almost total absence in academic and policy circles in Britain.[64] The strategic studies community addresses the issue rarely, and political debate between the two major parties on defence issues almost entirely excludes any discussion of nuclear matters. The present government is deeply committed to the current policy and the Labour Party seems utterly intent on avoiding any debate.

Even if Britain was reluctant to give up nuclear weapons, there is considerable scope to encourage a movement towards the comprehensive multilateral disarmament of weapons of mass destruction. These could include a commitment to no first

[62]See reference 42.

[63]An Evolving Nuclear Posture. Second Report of the Steering Committee, Project on Eliminating Weapons of Mass Destruction. The Henry L. Stimson Centre, Washington DC, 1995.

[64]A notable exception is the work of the British Pugwash Group. See, for example: C.R. Hill, R.S. Pease, R.E. Pierls and J Rotblat, *Does Britain Need Nuclear Weapons?* Pugwash, London, 1995.

use of nuclear weapons, support for rapid negotiation of nuclear-free zones, cessation of continual patrols of Trident submarines, a freeze on warhead development or modification and verifiable decreases in warhead and missile numbers. Increased support for the funding of non-proliferation treaty activities, especially in relation to verification, could be accompanied by much more vigorous technical support for existing activities. The completion of the Chemical Weapons Convention and the comprehensive upgrading of the Biological Weapons Convention should be foreign policy priorities, and a preparedness to enter negotiations on a Nuclear Weapons Convention would be desirable.

There is little prospect for such changes at present, although developments in international political opinion in the wake of the French tests may eventually put some pressure on the British government. Meanwhile, the deployment of the Trident system remains a slow-burning fuse. The UK Polaris missile system survived thirty years without use; with its Trident replacement deployed into a more uncertain and volatile world, the prospects are less good. It would be wise to seek alternative policies.

Chapter 10
Big Boats and Bigger Skimmers (2006)

10.1 From Cold War to Long War

In response to the attacks of 9/11 and in the context of a vigorous stance on international security, decisions being taken in the United States are setting in motion a potentially long-lasting military posture.[1] Although not on the same scale, similar decisions are being made in Britain concerning the development of a substantially enhanced expeditionary warfare capability focused on two large aircraft-carriers and the replacement of Britain's current Trident nuclear weapons system.

The Pentagon's Quadrennial Defense Review, published in February 2006, demonstrates a clear determination by the United States to maintain its current robust posture in relation to 'rogue' states and also to Al-Qaeda and its associates, with a strong emphasis on special forces, long-range strike and a variety of systems of force projection. A strong thread running through the document is a perceived requirement to take pre-emptive military action against rogue states and paramilitary opponents wherever they might be identified.

In parallel with these developments, a process of nuclear weapons modernization is now under way that involves maintaining core strategic forces for several decades to come, while re-establishing a range of development and production facilities for producing new nuclear warheads. The Minuteman III ICBM fleet, for example, is being upgraded to maintain it in flexible deployment for the next 30 years, and consideration is being given to an eventual replacement that will have a life of at least a further 30 years. It is likely that nuclear weapons research will be consolidated in one of the two current laboratory complexes, Lawrence Livermore or Los

---

[1] This text was first published as: "Big boats and bigger skimmers: determining Britain's role in the long war", *International Affairs,* 82, 4 (2006), pp. 651–665, John Wiley for Chatham House, (Royal Institute of International Affairs). The permission to republish this text was granted by Oxford University Press on 28 February 2018.

P. Rogers, *Paul Rogers: A Pioneer in Critical Security Analysis and Public Engagement*, Pioneers in Arts, Humanities, Science, Engineering, Practice 21,
https://doi.org/10.1007/978-3-319-95150-8_10

Alamos, with an emphasis on the development of small warheads intended for specific uses such as the destruction of hardened targets (see footnote 1).

The phrase that is gaining parlance to encapsulate the thinking behind this approach is 'the Long War', in conscious echo of the paradigm it is seen as replacing: the Cold War. Clinton's first CIA director, James Woolsey, talked in 1998 of having 'slain the dragon' of the old Soviet Union but seeing it replaced by 'a jungle full of snakes'.[2] The new US security mission is to tame the jungle and kill the snakes. Conventional forces will be configured substantially for this purpose while the United States will retain powerful and diverse nuclear arsenals.

Since the 9/11 attacks the United Kingdom has been the closest ally of the United States, and one of the most interesting questions about this new US security paradigm is whether future UK governments will remain committed to sharing it to the extent that the Blair government does. Particularly significant in this context are the two separate decisions on major weapons projects likely to be taken in the lifetime of this government, which will largely determine whether the UK defence posture over the next 30 years does embrace this paradigm. One of these planned decisions is well known – the replacement of the Trident nuclear system.[3] If Trident is replaced by a flexible system such as the current two-warhead configuration (strategic and sub-strategic), then this commits the UK to a nuclear posture that allows for usability, while limiting counter-proliferation policy largely to an outlook of 'do as we say, not as we do', with all the diplomatic and political difficulties that involves.

The second issue, which is far less widely noted, is the plan to replace the Royal Navy's three small *Invincible*-class aircraft-carriers with two very large 65,000-tonne *Queen Elizabeth*-class carriers in the next ten years. These will be the largest warships ever deployed by the Royal Navy, will carry the advanced (but extremely expensive) American F-35 multiple-role aircraft, and will give Britain a capacity to deploy worldwide that it has not had since the early 1970s.

If both programmes go ahead, then the costs of the Trident replacement and the new carriers and their aircraft will dominate defence procurement for the next two decades or more, and will take Britain firmly in the direction of the US outlook that appears to be largely embraced by the current Labour government. Significantly, it will make it very difficult for a future government to diverge from that approach unless it is to entertain a massive and costly programme of cancellations and reorientations.

The Trident replacement issue should not therefore be seen in isolation from the new carriers, for as a result of these two decisions, both to be taken in the near

---

[2]Testimony of R. James Woolsey at Senate Hearings, Committee on National Security, US House of Representatives, 12 Feb. 1998.

[3]Ken Booth and Frank Barnaby, eds, *The future of Britain's nuclear weapons,* Current Decisions Report, Oxford Research Group, Oxford, 2006.

future, the overall UK defence posture is likely to be set in stone for some decades. This article argues that the combination of the two will point the UK defence posture in the particular direction of maintaining the 'control' paradigm which is currently at the root of the US security outlook, making it less easy to consider a more radical reappraisal involving a move towards a 'sustainable security' outlook that might be a more appropriate response to that jungle full of snakes.

10.2 Trident Replacement

Because of the projected hull life of Britain's *Vanguard*-class nuclear missile submarines and the ten or more years it will take to develop a new system, the case is now being made that the decision on their replacement with a new national nuclear force should be made within the lifetime of the current parliament. There are arguments as to whether this is necessary, and some analysts believe that the decision can be postponed until sometime after 2010. At the same time there are strong indications that substantial investment is already being made in the UK's nuclear weapons design and development capabilities, to the extent that the decision to replace the existing nuclear weapons systems with newly designed weapons may, in effect if not in substance, have already been taken.

Britain's nuclear weapons are designed, developed and maintained at the Atomic Weapons Establishment at Aldermaston near Reading and at the nearby plant at Burghfield. Aldermaston is currently undergoing a massive upgrade of its facilities, so large that it has been described in an internal company newsletter as one of the largest construction projects ever undertaken in Britain, on a similar scale to the new £4.2 billion Terminal Five at London's Heathrow Airport. This does not seem an appropriate investment for a facility with the function of maintaining Britain's current nuclear weapons arsenal, given that there are only two warhead types, both carried on the *Vanguard* boats. At the very least it points markedly in the direction of a decision to replace Trident.

The main thermonuclear warhead currently carried by the Trident missile is of UK manufacture but is believed to be based closely on a US design, the W76 warhead. Britain's version may have a similar explosive power to the W76 – 100 kilotons, equivalent to 100,000 tons of conventional high explosive. In crude terms, a single Trident warhead is roughly eight times as powerful as the Hiroshima bomb that killed over 80,000 people on 6 August 1945. A number of the Trident warheads, possibly about one-sixth, are deployed as sub-strategic or tactical warheads, being configured to detonate as an 'unboosted primary' of around 1 kiloton or as a 'boosted primary' with an explosive yield similar to that of the older WE-177 tactical nuclear bomb, around 5–10 kilotons. Each submarine on patrol normally carries a mix of the single-warhead missiles armed with sub-strategic or tactical nuclear warheads and the much more powerful multiple-warhead missiles.[4]

In terms both of Britain's nuclear contribution to NATO and of independent targeting, Britain has long had a policy of nuclear first use. The NATO policy of the first use of nuclear weapons was not promoted widely in public, where all the emphasis was placed on nuclear weapons as an ultimate deterrent. Even so, the policy was made clear on relatively rare occasions, as in this example of evidence from the Ministry of Defence to a parliamentary select committee in 1988:

> The fundamental objective of maintaining the capability for selective sub-strategic use of theatre weapons is political – to demonstrate in advance that NATO has the capability and will to use nuclear weapons in a deliberate, politically-controlled way with the objective of restoring deterrence by inducing the aggressor to terminate his aggression and withdraw.[5]

Since the 1950s, Britain has deployed nuclear weapons on many occasions outside the immediate NATO area of western and southern Europe and the North Atlantic. These included the basing of RAF nuclear-capable strike aircraft in Cyprus in the 1960s and 1970s, regular detachments of V-bombers to RAF Tengah in Singapore in the mid-1960s, and the deployment of Scimitar and Buccaneer nuclear-capable strike aircraft on the Royal Navy's aircraft-carriers from 1962 to 1978.[6] Nuclear weapons were also carried on four taskforce ships during the Falklands/Malvinas War of 1982.[7]

The potential use of nuclear weapons away from the NATO/Warsaw Pact confrontation was referred to by the then Minister of Defence, Harold Macmillan, in 1955:

> the power of interdiction upon invading columns by nuclear weapons gives a new aspect altogether to strategy, both in the Middle East and the Far East. It affords a breathing space, an interval, a short but perhaps vital opportunity for the assembly, during the battle for air supremacy, of larger conventional forces than can normally be stationed in those areas.[8]

The idea of a small nuclear war was further expressed during the 1957 defence debate by one of his successors in the role, Duncan Sandys:

---

[4]The most consistent analysis of Britain's nuclear force is available periodically in the *Bulletin of the Atomic Scientists.* See Robert S. Norris and Hans M. Kristensen, 'British nuclear forces', *Bulletin of the Atomic Scientists* 61: 6, pp. 77–9, Nov–Dec. 2005. In relation to the role of sub-strategic Trident, the authors quote a Ministry of Defence official in 1996 describing a sub-strategic strike as 'the limited and highly selective use of nuclear weapons in a manner that fell demonstrably short of a strategic strike, but with a sufficient level of violence to convince an aggressor who had already miscalculated our resolve and attacked us that he should halt his aggression and withdraw or face the prospect of a devastating strategic strike'.

[5]House of Commons Select Committee on Foreign Affairs, *Third Report, 1987–88,* p. 35, para. 6.

[6]Paul Rogers, 'Sub-strategic Trident: a slow-burning fuse', *London Defence Studies* no. 34 (London: Centre for Defence Studies, 1996).

[7]Lawrence Freedman, *The official history of the Falklands campaign* (Abingdon: Routledge, 2005).

[8]*Hansard,* House of Commons, vol. 566, col. 2182, 2 March 1955.

> one must distinguish between major global war, involving a head-on clash between the great Powers, and minor conflicts which can be localised and which do not bring the great Powers into direct collision. Limited and localised acts of aggression, for example, by a satellite Communist State could, no doubt, be resisted with conventional arms, or, at worst, with tactical nuclear weapons, the use of which could be confined to the battle area.[9]

This historical context raises the question as to whether sub-strategic, or indeed strategic, Trident warheads might be used independently of NATO. Britain reserves this right, and one of the more detailed assessments of the range of options for sub-strategic Trident warheads was made in the authoritative military journal *International Defence Review* in 1994:

> At what might be called the 'upper end' of the usage spectrum, they could be used in a conflict involving large-scale forces (including British ground and air forces), such as the 1990–91 Gulf War, to reply to an enemy nuclear strike. Secondly, they could be used in a similar setting, but to reply to enemy use of weapons of mass destruction, such as bacteriological or chemical weapons, for which the British possess no like-for-like retaliatory capability. Thirdly, they could be used in a demonstrative role: i.e. aimed at a non-critical uninhabited area, with the message that if the country concerned continued on its present course of action, nuclear weapons would be aimed at a high-priority target. Finally, there is the punitive role, where a country has committed an act, despite specific warnings that to do so would incur a nuclear strike.[10]

It is worth noting that three of the four circumstances envisaged involve the first use of nuclear weapons by Britain.

Concern has been voiced in parliament that the government has not been sufficiently clear about the circumstances under which British nuclear weapons would be used in post-Cold War circumstances. For example, the House of Commons Defence Select Committee noted in 1998 its

> regret that there has been no restatement of nuclear policy since the speech of the then Secretary of State in 1993; the SDR [Strategic Defence Review] does not provide a new statement of the government's nuclear deterrent posture in the present strategic situation within which the sub-strategic role of Trident could be clarified. We recommend the clarification of both the UK's strategic and sub-strategic policy.[11]

The reality remains obscure. Indeed, prior to the 2003 invasion of Iraq, statements by the then Defence Secretary, Geoff Hoon, suggested that Britain would be prepared to use nuclear weapons in response to Iraqi use of chemical and biological weapons.

In short, Britain retains a flexible nuclear force with two warhead types and a targeting policy that allows for first use. It is just possible that a Trident replacement could take the form of a simple system with small numbers of large warheads intended solely for the purpose of deterring nuclear attack on Britain. This is frankly implausible in the context of the 50-year history of British nuclear weapons and

[9] *Hansard,* House of Commons, vol. 568, col. 1765, 16 April 1957.

[10] David Miller, 'Britain ponders single warhead option', *International Defence Review,* Sept. 1994, pp. 45–51.

[11] House of Commons Select Committee on Defence, *Eighth Report, 1997–98,* para. 152.

their previous deployment and targeting policies. If Trident is replaced, it is far wiser to assume that the new system will have at least as much flexibility as the current system. As such, Britain will aim to retain its nuclear status well into the 2050s and will not comply with Article VI of the Non-Proliferation Treaty, but will continue to argue against other states such as Iran developing their own systems. That is to say, in this key aspect of the UK defence posture there would be no change.

10.3 The New Aircraft-Carriers

Where there will be a major change is in the implementation of the decision to build the two large new aircraft-carriers that was announced in the Labour government's 1998 Strategic Defence Review. These are intended to replace the three small *Invincible*-class ships, and when they are fully operational with the F-35 by around 2020, the Royal Navy will acquire a global strike capability for the first time in well over 40 years. This is a very substantial change in the UK defence posture and should be seen in its historical context.

Back in the 1950s and 1960s, the Royal Navy had two large fleet carriers, *Eagle* and *Ark Royal*, and a handful of light carriers. Both of the 53,000-tonne ships had been begun in the later years of the Second World War but were completed in modernized form some years afterwards—*Eagle* in 1952 and *Ark Royal* in 1955. The eventual intention was to supplement them with two large modern carriers of similar size, CVA-1 and CVA-2, that would reflect the developments in carrier design of the postwar era.

After the Suez crisis of 1956 and the subsequent decision to withdraw from east of Suez, the programme was cancelled in 1966, eventually leaving the Royal Navy with the two existing fleet carriers and three light carriers that were subsequently converted into commando carriers, *Albion, Bulwark* and *Hermes.* This still left the Royal Navy with quite a substantial carrier force, but only until the 1970s. *Eagle* was scrapped in 1972 and *Ark Royal,* the navy's last fleet carrier, went in 1978. *Albion* and *Bulwark* were also scrapped, in 1973 and 1981 respectively; *Hermes* survived into the 1980s and was eventually sold to India.

The intention behind these decisions was to allocate airborne maritime air defence, strike and reconnaissance functions to the Royal Air Force, saving the considerable costs of maintaining a force of large carriers with their specialized aircraft and all their support requirements. As a result of this decision the Royal Navy lost its carrier-based global reach capability.

In the Cold War context, though, the Royal Navy still required a substantial anti-submarine warfare (ASW) capability, and this was to be met partly through the construction of a new class of ASW ships known as anti-submarine command cruisers or 'through-deck cruisers'. These were effectively small helicopter carriers, although the use of that term was largely avoided during their development for political reasons. The first of the new 18,500-tonne *Invincible*-class ships was ordered in 1973 and entered service in 1980, to be followed by *Illustrious* in 1982 and finally by a new *Ark Royal.* While these were initially intended as helicopter carriers, the development of the Sea Harrier and the ski jump meant that they also had a limited capability to carry fixed-wing aircraft, as did *Hermes,* which had meanwhile been refitted with a ski jump as a temporary measure. Both *Invincible* and *Hermes* were deployed in the Falklands/Malvinas War of 1982.

Since the mid-1980s it has been theoretically possible for the Royal Navy to deploy two small carrier task groups based on the *Invincible*-class ships, although only one has normally been available, carrying up to ten helicopters and eight Harriers. The 1998 decision to build two new fleet carriers will therefore hugely increase capability, returning to the Royal Navy long-range capabilities last available in the 1970s.

Each of the new CVF or *Queen Elizabeth-class* ships (the second being *Prince of Wales)* will displace up to 65,000 tonnes, making them by far the largest warships ever to see service in the Royal Navy, much larger even than the battleships of the Second World War era. Each will have an air wing of around 55 aircraft, including 42 STOVL (short take-off vertical landing) multi-role aircraft and a range of antisubmarine, airborne early warning and rescue helicopters. Final approval for the production of the carriers is expected by summer 2007, with completion of HMS *Queen Elizabeth* being scheduled for 2014, already two years later than originally planned. The carriers are being designed to have a hull life of up to 50 years, meaning that they could remain operational until the mid-2060s.[12]

The aircraft intended for the new carriers, the F-35, is also known as the Joint Strike Fighter (JSF) or, in UK parlance, the Joint Combat Aircraft (JCA). This is one of the world's most expensive military projects ever undertaken, currently costed at $256 billion (£144 billion).[13] Britain intends to buy up to 150 planes at a cost of $104 million (£59 million) each, to equip the new carriers as well as land-based RAF squadrons. Apart from the United States and Britain, six other countries are involved in the programme, although the British order is by far the largest. Some of the other partners are already questioning costs and viability; Italy is considering cancellation, following the recent election of Romano Prodi as prime minister of a centre-left coalition,[14] and Norway is also reconsidering its involvement.[15]

Although Britain remains committed to the project, there are concerns at cost overruns and delays in the project, not least because of experience with the last major aircraft project in which the country became involved, the Eurofighter or

[12]'CVF—Royal Navy future aircraft-carrier, United Kingdom', *Naval Technology,* http://www.naval-technology.com/projects/cvf/, accessed 1 May 2006. In submariners' parlance, submarines tend to be termed 'boats' whereas surface warships such as aircraft-carriers are 'skimmers'.

[13]Leslie Wayne, 'US and allies' fighter jet program hits turbulence', *International Herald Tribune,* 17 March 2006, p. 11.

[14]Tom Kington, 'Will Italy cancel JSF buy?', *Defense News,* 17 April 2006, p. 1.

[15]Douglas Barrie and Robert Wall, 'Nordic air battle', *Aviation Week and Space Technology,* 8 May 2006, p. 32.

Typhoon. This was conceived more than a quarter of a century ago during the Cold War, when a new plane was required to specialize in combat with Soviet MiG-29s and Sukhoi-27s. Britain took the largest share of this multinational project, in which Germany, Italy and Spain were also involved, and which acquired such momentum that it survived the ending of the Cold War, despite that specific conflict being the major reason for its instigation. The project has since been beset by massive cost inflation and continual technical difficulties. Estimated costs for the UK have gone up from £7 billion to £19 billion and entry into service has been repeatedly delayed.[16] Known in some circles as the world's first supersonic white elephant, in 1997 it was described by Alan Clark, former minister of state for defence, as 'essentially flawed and out of date'. Commenting on its role in job creation, he said, 'we must find a less extravagant way of paying people to make buckets with holes in them'.[17] Intended to start replacing the flawed interceptor variant of the ageing Tornado in the mid-1990s, the Typhoon is still not yet in full squadron service, the operational employment date having now been shifted from 2006 to 2009.

Although the F-35 is at a much earlier stage of development, major controversies have already arisen. There is serious doubt over an earlier plan to have two sources of engines for the plane, with Rolls-Royce potentially losing out in the process. There have also been persistent concerns that aspects of the F-35 technologies would not be made available to partner nations, making it very difficult for countries such as Britain to develop their own upgrades for the plane.

In spite of these problems, the current intention is to continue with both the aircraft-carriers and the F-35. Indeed, France is investing in the CVF design as it may build a similar ship for its own navy. France's current aircraft-carrier, the nuclear-powered *Charles de Gaulle*, has been beset by engineering problems and the French government is leaning heavily towards a conventionally powered ship as its second carrier. Such a French connection would make it more difficult to cancel Britain's CVF project.

If the carrier project does go ahead, as seems likely at the time of writing, past experience suggests that there will be substantial delays in completing the ships and in acquiring the F-35 aircraft, with extensive cost overruns. Even so, the end result is likely to be that, by 2020, Britain will be deploying two large carriers with a potential for global reach. While these will not be as powerful as the US navy's 100,000-tonne *Nimitz*-class carriers, they will greatly exceed the largely symbolic roles that the *Invincible*-class ships have played in joint operations with the United States such as the first and second Iraq wars. The carriers, with their escorts and support ships, together with the Trident replacement, will dominate the Royal Navy's procurement budget over the next two decades and much of its overall budget for some decades after that.

[16]National Audit Office, *Ministry of Defence Major Projects Report: project summary sheets,* p. 119, http://www.nao.org.uk, last accessed 1 May 2006.

[17]*Hansard,* House of Commons, vol. 297, col. 855, 9 July 1997.

While most public attention is focused on the arguments for and against replacing Trident, there is a strong case for looking at the nuclear force alongside the planned carriers, for two quite different reasons. One is that they will point the overall UK defence posture in one particular direction. The second is that that direction involves enhancing and expanding existing capabilities in a manner that parallels US developments, albeit on a smaller scale, rather than allowing for a serious reassessment of Britain's defence policy.

10.4 The United States from Cold War to Long War

Following the ending of the Cold War and the immediate experience of the 1990–91 Iraq crisis and war, the 1990s saw a series of changes to the composition of the US armed forces that amounted to a reorientation towards taming the jungle described so vividly by James Woolsey in 1993. There was a substantial drawdown in nuclear forces, especially tactical nuclear weapons, and the US air force invested more heavily in long-range conventional strike and stand-off capabilities – 'global reach' was the order of the day. The US army cut back on its heavy armour, especially those forces based in Europe, and put relatively more resources into special operations forces, while the US navy reduced its major commitments to anti-submarine warfare, now that the successor states of the Soviet Union were deploying very small forces compared with those of the Cold War era.

Significantly, there were two areas in addition to special forces where the cuts were minimal. The US navy retained almost all of its aircraft-carrier battle groups, maintaining the capability to project force on a global scale, and the US Marine Corps, with its emphasis on expeditionary warfare, had the smallest personnel cuts of any of the four branches of the armed forces. Furthermore, while there had been substantial cuts in nuclear forces, one important exception was the modification of the standard B61 tactical nuclear weapon to reconfigure it as an earth-penetrating warhead, the B61-11, with a hardened casing and modified fusing. This gave the US air force an enhanced capability for destroying hardened targets such as command centres or underground bunkers housing chemical or biological weapons.

In overall terms, the United States ended the 1990s with leaner military forces, having cut personnel on average by around 25 per cent while preserving those elements more likely to be used in responding to regional crises that fell far short of a global threat. The election of George W. Bush to the presidency in November 2000 did not have a major effect on the defence posture, although it brought in an administration with much more unilateralist tendencies. These included firm opposition to the Comprehensive Test Ban Treaty, a proposal to withdraw from the Anti-Ballistic Missile Treaty, opposition to the strengthening of the 1972 Biological and Toxin Weapons Convention, and opposition to the proposed International Criminal Court and to planned negotiations to control the introduction of weapons into space. Most surprising to European governments was the withdrawal from the Kyoto climate change accords, but this fitted an administration outlook that was deeply suspicious of the handicapping effects of international agreements on US foreign and security policy—an outlook expressed succinctly by Charles Krauthammer in July 2001, a few weeks before the 9/11 attacks:

> Multipolarity yes, when there is no alternative. But not when there is. Not when we have the unique imbalance of power that we enjoy today—and that has given the international system a stability and essential tranquillity that it had not known for at least a century. The international environment is far more likely to enjoy peace and a single hegemon. Moreover we are not just any hegemon. We run a uniquely benign imperium.[18]

The attacks on New York and Washington in September 2001 were deeply shocking at any level, but particularly so for an administration that was so convinced of its strength. The response was vigorous in the extreme and primarily military in nature. The Taleban regime in Afghanistan was terminated within three months, and this action was followed by the January 2002 *State of the Union* address which formulated the concept of an 'axis of evil' of states that supported terrorism and sought weapons of mass destruction. The early pre-emption of suspected threats to US security, even if they were entire regimes, was subsumed into the response to 9/11, and early in 2003 the invasion of Iraq and the deposing of Saddam Hussein ensued.

If the early months of 2002 represented the high point of the global war on terror, the situation has changed drastically in the past four years. Afghanistan is not making the transition to peace. Instead, warlordism, expanding illegal drug production and a marked Taliban revival are leading to increased insecurity across much of the country.[19] In Iraq, the United States is mired in an insurgency that has cost at least 40,000 civilian lives and is now developing into a bitter civil war.[20] Reconstruction of the Iraqi economy has been a signal failure, and the United States itself has lost 2,400 military personnel killed and 18,000 injured.[21] Tensions are rising between Iran and the United States, with the possibility of US or Israeli air strikes against Iranian nuclear facilities no longer regarded as far-fetched.

In the pursuit of the global war on terror by the United States and its allies since 9/11, around 100,000 people have been detained without trial as terror suspects, some of them for four years or more. Torture, prisoner abuse and rendition have become common practice and tough new legislation has been enacted in the United

---

[18]Charles Krauthammer, 'The Bush Doctrine: ABM, Kyoto and the new American unilateralism', *Weekly Standard* (Washington DC), 4 June 2001, pp. 5–7.

[19]Carlotta Gall, 'Taliban power creeps back in Afghan south', *International Herald Tribune*, 4 May 2006, p. 2.

[20]Baseline figures for Iraqi civilian casualties are published by Iraq Body Count: www.iraqbodycount.net/, last accessed 7 May 2006. These are drawn from detailed media reports, and actual figures, including the many casualties not recorded in media outlets, are likely to be substantially higher.

[21]US and other coalition casualties are available from Iraq Coalition Casualties: last accessed 7 May 2006. In addition to these figures, more than 20,000 US military personnel have been evacuated to the United States because of non-combat injuries or physical or mental illness.

States, Britain and other coalition countries. It is perhaps surprising, therefore, that the Al-Qaeda movement has not only survived but is expanding its influence. In the five years since 9/11, the movement has achieved a far greater level of activity and support than in the previous five years. There have been attacks in Istanbul, Jakarta, Bali, Riyadh, Casablanca, Mombasa, Sharm al Shaikh, Taba, Karachi, London, Madrid and many other cities, with attempted attacks in Paris, Rome, Los Angeles and Singapore, among others – and this list excludes the huge array of attacks in Iraq and Afghanistan.

Most remarkable of all has been the manner in which Iraq has begun to evolve into a new jihadist combat training zone, attracting young radicals from across North Africa and south-west Asia. Just as opposition to the Soviet Union in the 1980s and the Northern Alliance in the 1990s made Afghanistan a training zone for a generation of jihadists, so Iraq is increasingly serving that function now. Some of the key aims of the Al-Qaeda movement, such as defeating the 'near enemy' of regional elite regimes and evicting the 'far enemy' of the United States from the region, involve guerrilla combat and other forms of asymmetric warfare. They are projects measured in decades rather than years and in such a context, the benefit arising from the availability of Iraq as a long-term training environment is difficult to overemphasize. What gives Iraq its real value is that the training in combat of a new generation of jihadists is being conducted directly against the forces of the otherwise overwhelmingly powerful 'far enemy' and is done so in an urban environment, not the largely dispersed rural environment of Afghanistan in which training was conducted in the 1980s and 1990s. This is an unexpected yet wholly welcome gift to the Al-Qaeda movement with potential benefits that could last for decades.

Under such circumstances – declining security in Afghanistan, failure in Iraq and the persistent activities of Al-Qaeda, it might be expected that there would be a questioning of US policy and the investigation of alternative approaches. Given the increasing domestic unpopularity of President George W. Bush and the problems of the war in Iraq, there would even appear to be sound reasons of domestic politics for such a rethink. So far, however, there is no evidence of this. Instead, there is a consolidation of the war on terror into the Long War, with a belief that the United States is now engaged in a long conflict that is fundamentally a matter of military confrontation to be resolved primarily by the use of force.

This outlook was encapsulated in the new Quadrennial Defense Review (QDR) published in February 2006,[22] and the defence budget for fiscal year 2007, published shortly afterwards.[23] The QDR system was mandated by Congress towards the end of the Clinton presidency and the 2006 review was the third in a series. The second had followed closely on the 9/11 attacks and was an early reaction to the new security dynamic, whereas the 2006 QDR looks forward in a

[22]Available at, last accessed 6 May 2006.

[23]Summarized at, last accessed 6 May 2006.

more articulate way. Taken together, the QDR and the FY 2007 defence budget show a strong reinforcement of existing policies, not a reassessment. The budget involves a spending increase of almost 7 per cent, taking it to $439.3 billion, around 45 per cent more than the budget inherited from Clinton in 2001 and almost as large as the combined military expenditure of all the other countries in the world put together.

Two core issues inform the QDR: the need for the United States to remain the world's pre-eminent state with the most powerful military forces, and the insistence on seeing the war on terror as a long-term matter. Many of the major existing military programmes survive, but several new features stand out: a marked emphasis on special forces,[24] the development of a more substantial long-range airborne strike capability, an increased capability against biological and nuclear weapons, and the ability to act against paramilitary groups anywhere in the world, whether or not that infringes the sovereignty of a particular state.

An example of the last strand has been the deployment by US Special Operations Command (SOCOM) of teams of specialists in US embassies in 20 countries. Originally termed 'Operational Control Elements', these are now described by the more anodyne phrase 'Military Liaison Elements'. These units have considerable freedom of action, not least in taking action against paramilitary groups that may be considered threatening to the United States. Moreover, whereas formerly such a unit wishing to carry out an operation in any country had to obtain approval from the US ambassador there, the Department of Defense has persuaded the Bush administration to allow these units to conduct operations while merely informing the relevant ambassador.[25]

Such increasingly vigorous policies are paralleled by an expansion in the US military's capability in unmanned combat aerial vehicles – drones such as the Predator, equipped with Hellfire air-to-surface missiles, that can be used for remote operations against paramilitary units in countries whether or not US forces are operating openly there. These operations have already been experienced in Yemen and Pakistan and many more are anticipated.

Overall, the QDR and the FY 2007 defence budget reinforce existing trends, the underlying feature being strong reliance on military force as the primary factor in US security. The specific examples, such as the enhancement of SOCOM, form part of a wider process of maintaining the force projection capabilities embodied in the navy's carrier battle groups, the Marine Corps' amphibious forces and the air force's long-range strike capabilities, while also ensuring that the United States modernizes its nuclear forces. The fundamental principle is that the appropriate response to a fragile, uncertain world of diverse threats is one of early and substantial recourse to military action, extending to pre-emption when necessary.

---

[24]David Tucker and Christopher J. Lamb, 'Restructuring Special Operations forces for emerging threats', *Strategic Forum,* no. 219 (Washington DC: Institute for National Strategic Studies, National Defense University, Jan. 2006).

[25]Ann Scott Tyson, 'New plans foresee fighting terrorism beyond war zones', *Washington Post,* 23 April 2006, p. 01.

10.5 The UK Connection: Convergence or Difference?

The British defence budget is small compared with that of the United States. In personnel terms, for example, the US Marine Corps is about three times the size of the entire Royal Navy, and US special forces alone are nearly half the size of the British Army. Even so, Britain's role alongside the United States in the latter's global war on terror has been politically significant. Tony Blair has been regarded as President Bush's closest ally, not least in terms of troop deployments to Afghanistan and Iraq, and the persistent commitment of the British government to the United States has been valuable in Washington's campaign to present its war on terror as a properly multinational endeavour. Many other countries have played significant roles, especially in Iraq, where Italy, Poland, Ukraine and others have had substantial deployments, and Afghanistan, where NATO is now responsible for the International Security Assistance Force; but Britain has been the most dedicated ally. Over the past five years, accordingly, the transatlantic partnership has been considerably strengthened, and now embraces a British commitment to working with the United States in matters of international security that betrays scarcely a shred of criticism of the conduct of the war on terror so far, or of the US defence posture that is now emerging in order to fight the Long War.

However, there are strong arguments for suggesting that the development of alternative strategies to those of the United States would make sound political sense, both in the short term and in the coming decades. The war on terror is configured against both states and sub-state movements, but the experience at both levels is unsettling. Neither Iraq nor Afghanistan shows a prospect of achieving the pro-American liberal market success that was anticipated, and if a military confrontation with Iran were to occur, then the consequences could well be long term instability and violence involving US forces to the east as well as the west of the Persian Gulf.

Meanwhile, the Al-Qaeda movement seeks to prosecute its many aims. In the next two to three decades these involve the eviction of western military forces from the region and the termination of the House of Saud, seen as a corrupt, elitist and unacceptably pro-western Keeper of the Two Holy Places. The movement also seeks the termination of other regimes that it sees as elitist and pro-western, such as those in Egypt, Jordan and the emirates, together with the successful creation of a strong Palestinian state. Further afield it supports local movements such as the separatists in southern Thailand and the paramilitary movements in Kashmir. In the longer term, over half a century or more, Al-Qaeda seeks the establishment of a properly Islamist caliphate.

While these aims, and their radical implementation, attract the support of only a minority among Islamic communities and others worldwide, the manner of the conduct of the war on terror by the United States and its allies has been deeply and persistently counterproductive to the opponents of the movement and its programme. It has systematically aided recruitment to the Al-Qaeda cause while fostering increasing opposition to a sustained US presence in Iraq. Across the region, there is a dominant view that US policy has little or nothing to do with the spread of democracy and human rights and much more to do with control of the region's oil reserves and unrestrained support for Israel.

It is possible to consider the development of an alternative approach that seeks to undermine support for the Al-Qaeda movement and its associates. This would involve a sustained high-level commitment to a just and lasting settlement of the Israeli/Palestinian confrontation, a managed withdrawal from Iraq, and promotion of human rights and greater socio-economic justice across the Middle East. Combined with promotion of the rule of law rather than a dominantly punitive approach to paramilitary movements, this would progressively undermine the support base of radical jihadist movements such as Al-Qaeda. Such a change of approach is highly unlikely in the short term, at least from the United States. Elsewhere, however, an outlook is beginning to emerge in relation to global security trends that embraces a very different approach from the control paradigm of the Long War: one that has been called a sustainable security paradigm.[26]

Three issues are beginning to be seen as dominating international affairs at the global level in the next three decades or so. One is the deepening socio-economic divide that has been evolving for several decades and now sees a transglobal elite of well over a billion people benefiting greatly while the majority of the remaining five billion are marginalized. That majority, though, is far more literate, educated and aware of its marginalization than it has been in former times, and this leads to responses that range from the political changes currently affecting much of South America through to the emergence of new radical social movements such as the Maoist revolt in Nepal and the reinvigoration of Naxa movements in India. While many Al-Qaeda operatives are highly educated and far from poor, a large measure of the overall support for this and similar movements comes from many millions of people who certainly do see themselves as marginalized from elite power bases, whether those be regional or global.

The second issue, increasingly interacting with socio-economic divisions, is the evolution of environmental constraints as human activity impinges on the global ecosystem. Among the many impacts of climate change, including sea-level rises and the increasing intensity of storms, the most substantial security impact is likely to be the progressive drying out of the tropics and sub-tropics as world rainfall patterns shift from the tropical land masses to the oceans and the polar regions.[27] The effects of such changes on the ecological carrying capacity and hence crop productivity of tropical agriculture, and in turn on marginalized societies in these regions, should be expected to include greatly increased migratory pressures, the development of new radical social movements, and considerable social and political unrest.

---

[26]This wider approach to security is explored in: Chris Abbott, Paul *Rogers and John Sloboda,* Global responses to global threats: sustainable security for the 21st century (Oxford: Oxford Research Group, June 2006).

[27]One of the first reports of this trend was: David Rind, 'Drying out the tropics', *New Scientist, 6* May 1995, pp. 36–40.

A related question is that of the location of certain natural resources, especially the most readily usable fossil fuels such as oil and gas. With close to two-thirds of the world's oil reserves and a substantial minority of the world's gas reserves located around the Persian Gulf, and with China, India and the United States having joined western Europe and Japan as major oil importers, the risk of intense competition and conflict is already evident.

Finally, in a constrained and divided world, the proliferation of techniques of asymmetric warfare are counterbalancing the determination of the United States and others to maintain control without addressing the underlying roots of opposition.[28] There is little sign that the control paradigm will work, and yet even as the failure of the post-9/11 responses provides evidence of this, renewed efforts are made to prepare for the Long War.

10.6 The Carriers and the Trident Successor

One of the ironies of the British Labour government's decisions on the new aircraft-carriers and Trident replacement is that it has otherwise tended to adopt a relatively liberal attitude to global issues, at times even suggesting an engagement with the concept of sustainable security. Since coming to power in 1997, the Labour government has substantially increased Britain's international development budget, has sought to direct aid particularly at the poorest communities, and has successfully placed debt relief on the agendas of successive G8 meetings. Its commitment to trade reform to assist the poorest countries has been less evident, but it has been reasonably strong on issues of climate change, even if the actual achievements in curbing greenhouse gas emissions have been modest at best.

The problem is that there has been a marked discontinuity between these policy developments, which do exhibit a degree of global awareness, and the more immediate and very vigorous responses to the 9/11 attacks and subsequent incidents. The impact of those attacks, especially of the London bombings of July 2005 – and especially, it seems, on Mr. Blair – has been to bring about a hardening of the old paradigm, involving resolute support for the Bush administration and its pursuit of the war on terror, sometimes embracing an almost Manichaean assumption of good versus evil.

In one sense, the decisions due to be taken on the building of the two new aircraft-carriers and the replacement of Trident are coincidences. They could have come at any time, but they happen to fall due when defence thinking in Britain, at least around the office of Prime Minister, is hugely influenced by perceived threats from that snake-infested jungle. It is certainly possible that political changes in

[28]Some of the underlying trends in military power projection are explored in Andrew Dorman, Mike Smith and Matthew Uttley, *The changing face of military power: joint warfare in an expeditionary era* (Basingstoke: Palgrave, 2002).

Britain over the next decade or more might bring to power Conservative governments or coalitions that take a very similar view and might even be stronger in their support for nuclear forces, not least in their ability to support the kind of conventional global reach embodied in the new aircraft-carrier programme. Alternatively, it might be that another five or so years of failure and setbacks in the global war on terror will lead to a fundamental reassessment of defence and security policy, embracing ideas of sustainable security; but there is little sign of that so far. Meanwhile it is in the more immediate future that the Trident-replacement and carrier decisions will be taken, and once taken they will make any such reassessment much more difficult. That is an outcome that could have implications for the British outlook on international security for much of the new century.

Part V
Selected Texts on Global Political Violence

Chapter 11
Responding to the World Trade Centre and Pentagon Attacks

The group responsible for the attacks has engaged in detailed planning over many months and has substantial numbers of supporters with total dedication to its aims. It does not appear to be strongly hierarchical, has a cellular international network and may be in loose coalition with like-minded groups.

It should be assumed to be operating in the context of a long-term strategy. And it should also be assumed to have the near-term capability for further attacks, either using hi-jacking or some other method(s) with equivalent or greater effect.

Further actions may be attempted in the near future, but, given the devastating effects of the World Trade Centre attacks, the group may consider this counter-productive to its interests, serving mainly to further encourage the substantial coalition of states that is ranged against it.

The specific aims of the attacks in New York and Washington were to have an immediate and lasting effect on US financial, military and political centres. The more general aim was to demonstrate some innate vulnerabilities in the world's most powerful country that is otherwise regarded as unassailable.

The group was part of a wider network that draws much of its strength from an endemic anti-US view that is particularly strong in many parts of the Middle East and south-west Asia. This has been heightened, in part by what is perceived to be the "aggressive containment" of Iraq, the overt control of the Persian Gulf, and, in particular, by the continual violence in Israel/Palestine and the widely perceived US support for an unusually hard-line Israeli government.

A core aspect of the current situation is that the group responsible for the attacks needs a strong US counter-reaction. *Indeed this should be recognised as one of the prime motivations for the attacks.* The group will have prepared for this and will have dispersed its assets and key personnel.

From its perspective, the most desirable US response would be widespread military action, both immediate and sustained, against training, logistical and other anti-US paramilitary facilities in several countries, together with direct attacks on the Kabul and Baghdad regimes.

P. Rogers, *Paul Rogers: A Pioneer in Critical Security Analysis and Public Engagement*, Pioneers in Arts, Humanities, Science, Engineering, Practice 21,
https://doi.org/10.1007/978-3-319-95150-8_11

If the US takes any such action it will be precisely what the group wants – indeed the stronger the action the better, especially if it involves civilian casualties and stretches over a number of months. In the view of the group, such action will have many desired effects, Among these, it will:

- Weaken the wide-ranging pro-US international coalition,
- weaken the position of the more moderate elements of the Kabul regime,
- strengthen opposition to the US in countries such as Egypt and Jordan,
- aid the destabilisation of the Pakistani government, making a change of (nuclear capable) government more feasible, and
- above all, enable the group to gain more support and recruits.

The group should also be expected to respond to such action with further paramilitary attacks in the US or against trans-national US interests or its allies. It should be anticipated that such a response would be at least as devastating as the recent attacks.

It is less likely to stage immediate attacks in the absence of a major US military response, as these would further isolate it.

Thus, vigorous military action by the US, on its own or in coalition, will be counter-productive, whatever the intense and understandable domestic pressures for such action.

Given the extent of the devastation and human suffering, support for the United States among its allies is far-reaching, and extends to a remarkable range of states.

Given this, the immediate response should be to:

- develop and extend this coalition,
- base all actions on the rule of law, and
- put every effort into bringing the perpetrators to justice.

The longer-term response should be to:

- Greatly improve intelligence and cooperation,
- Substantially strengthen international anti-terrorism agreements, and
- Analyse, understand and then seek to reduce the bitter and deep-seated antagonism to the United States in South West Asia and the Middle East from which these actions and groups have arisen.

The group responsible welcomes and seeks military confrontation. It is more fearful of being brought to trial, a process that is likely to weaken it far more than direct military action.

In the longer term, a systematic strengthening of multilateral co-operation against political violence is essential, and is more possible now than at any time in the past two decades.

This, combined with a substantive attempt to address the deep-seated antagonisms in the region, would be the most effective way of countering the further development and actions of such radical paramilitary organisations.

However, the Bush administration includes key advisors who should be expected to take a hard-line position, recognising few aspects of the foregoing analysis. At the same time a hard-line Israeli government is already using the recent events to engage in a more forceful action to control the *intifada*.

This is a volatile combination that should be expected to lead to further instability and violence. It is highly unlikely to bring paramilitary movements under control – indeed it is more likely to lead to their further development and enhancement.

It is particularly important for states close to the United States to endeavour to influence the Bush administration away from military action.

17 September 2001

Chapter 12
Political Violence and Global Order (2002)

A frequently quoted definition of terrorism is 'the threat of violence and the use of fear to coerce, persuade and gain public attention'.[1] Assigning political motives to terrorism[2] may be considered to provide a legitimacy to acts that public authorities would normally want to deny, but the political context is accepted by most analysts and is accurately reflected in Wardlaw's definition of political terrorism as:

> the use, or threat of use, of violence by an individual or a group, whether acting for or in opposition to established authority, when such an act is designed to create extreme anxiety and/or fear-inducing effects in a target group larger than the immediate victims with the purpose of coercing that group into acceding to the political demands of the perpetrators.[3]

An important aspect of such a definition is that it recognizes that terrorism and political violence are not the prerogative of substate groups. 'Agitational' or 'revolutionary' terrorism may be employed by such groups, but 'enforcement' or 'establishment' terrorism is the province of the state.

This elaboration is essential because state authorities concentrate almost exclusively on threats to their own power, whereas most terrorism is actually conducted by states against their own populations. The attacks on September 11 were widely regarded as unique, not least because of the considerable loss of life, but the use of mass-casualty political violence in the second half of the twentieth century was much more commonly an aspect of state control, and continues to be so.

There have been many examples of massacres of political opponents and non-politically active civilians across the world, from Indonesia to Iraq and from Chile to Guatemala, with many individual periods of violence far exceeding the losses of the September 11 atrocities. The use of chemical weapons of mass

---

[1]This text was first published as: "Political Violence and Global Order", Chapter in: Ken Booth and Tim Dunne, *Worlds in Collision: Terror and the Future of Global Order*, (2002) pp. 215–225, Palgrave/St. Martin's Press. The copyright was retained by the authors of individual chapters of the original book.
[2]Quoted in Sean Anderson and Stephen Sloan, *Historical Dictionary of Terrorism* (Metuchen, NJ, and London: The Scarecrow Press, 1995). This includes a helpful discussion of the conceptualization and definition of terrorism.
[3]Ibid.

P. Rogers, *Paul Rogers: A Pioneer in Critical Security Analysis and Public Engagement*, Pioneers in Arts, Humanities, Science, Engineering, Practice 21,
https://doi.org/10.1007/978-3-319-95150-8_12

destruction by Iraq against the Kurdish inhabitants of Halabjah in March 1988, killing some 5000 people, is a particular example. It was not greatly condemned by Western states, as Iraq was then regarded as a bulwark against Iran.

The use of terror continues in many parts of the world, including in countries that maintain support from Western democracies. Methods include summary detention without trial, torture and the use of paramilitary death squads and counter-insurgency military action against civilian populations. On any scale of measurement, terrorism is primarily a function of the state, even if the greatest attention is currently given to substate terrorism against Western states.

Even so, this chapter concentrates primarily on substate actors, but it will be argued that the changing 'drivers' or parameters of international security mean that it will be the interaction of state power with anti-state or 'anti-elite' action that will be particularly significant. While my focus in this chapter will, in part, be on the attacks of September 11 and their aftermath and consequences, my principal concern is with the wider issues of international security which, it will be argued, are likely to determine the nature and extent of conflict, including political violence, in the next two to three decades.

My intention is therefore to examine the main 'drivers' of conflict and insecurity, to contrast these with the current Western security paradigm, and to question whether that paradigm is sustainable. I analyse the prospect that advanced urban industrial states are sufficiently vulnerable to political violence and asymmetric warfare that it is in their own self-interest to address root causes of insecurity, and ask whether the impact of September 11 will help or hinder that process.

12.1 Trends in Conflict and Insecurity

The three dominant drivers of conflict and insecurity over the next few decades are likely to be socioeconomic divisions, environmental constraints and proliferation of military technologies. The latter results largely from the impetus and momentum of the Cold War years and includes the spread of weapons of mass destruction, ballistic and cruise missiles, area-impact munitions and the capability of paramilitary groups to target nodes of economic or political power using relatively crude devices and tactics.

In terms of socioeconomic divisions, the global trend of recent decades has been for an elite of about 1 billion to move rapidly ahead of the remaining 5 billion, with the ratio of inequality between the richest and poorest fifths of the global population nearly doubling. The gap widened markedly during the 1980s and 1990s and demographic and economic trends suggest a further widening. Elite communities are concentrated in the North Atlantic community and some West Pacific states, but with significant elites in many poor countries, and marginalized minorities in some rich countries.

A form of economic apartheid is emerging as the elite surges ahead of the rest and the globalized free market delivers patchy economic growth but fails to ensure any degree of economic justice. This process is happening in the context of significant improvements in primary education, literacy and communications, resulting in many more people becoming aware of their very marginalization. This, in turn, leads to the phenomenon of the 'knowledgeable poor' and a potential revolution of frustrated expectations.

Frequent results include substantial and increasing pressure to migrate, endemic problems of criminality and, in many parts of the world, anti-elite insurgencies and rebellions often stemming from the development of radical social movements. Among many examples are the Zapatista rebellion in Mexico, the Shining Path in Peru and FARC in Colombia, a bitter anti-authority revolt and civil war in Algeria, a Maoist rebellion in Nepal and protracted insurgencies in several parts of the Middle East and southern and Southeast Asia. Such trends should be expected to continue and increase in impact.

Environmental limitations are primarily felt, so far, in terms of resource conflicts. Smaller-scale examples include conflicts in Africa involving cobalt, diamonds, tantalum and rock phosphates, but the most notable example stems from the remarkable concentration of two-thirds of the world's proven oil reserves in the Persian Gulf. With the United States experiencing an increasing dependence on imported oil, now running at over 60 per cent of consumption compared with 10 per cent 30 years ago, a key component of US security policy has been an extensive military basing in the region, including the Fifth Fleet in the Persian Gulf and a major air force presence in Saudi Arabia. In turn, one of the major motivations of the al-Qaeda network, with considerable local support, has been an increasingly violent opposition to this presence.

In the longer term, though, certain global environmental impacts are likely to be even more significant. In the 1980s, the first evidence of global environmental change effected by human activity became apparent with the developing problem of ozone depletion caused by CFC pollutants. That was a relatively simple matter to address and some action has been taken, although not sufficient to prevent some serious environmental problems. Much more important is the phenomenon of climate change caused primarily by release of carbon gases due to excessive combustion of fossil fuels.

Until recently it was assumed that climate change would have its primary impact in temperate latitudes, causing substantial problems; not least in the form of severe weather events. These effects would at least be experienced mainly by wealthy countries that would best be able to cope. What is becoming clear is that climate change is likely to have profound effects on the tropical and subtropical land masses where the great majority of the world's people live.[4] A progressive 'drying out' of the tropics is thought likely, with profound effects on the ecological carrying capacity of some of the world's most fertile croplands. The effects of this would be to cause regional food deficits that would far outstrip even the major famines of the

[4]David Rind, 'Drying Out the Tropics', *New Scientist*, May 6, 1995.

past century, putting an immense strain on socio-political systems and greatly exacerbating migratory pressures.

Overall, the combination of socioeconomic divisions and environmental constraints implies enhanced problems of environmental conflict, insecurity stemming from mass migration and, above all, anti-elite insurgencies, some of them transnational in nature and effect. The results of such developments are inherently unpredictable, but what should be expected is that radical and extreme social movements are likely to develop that are frequently anti-elite in nature and draw substantial support from the marginalized majorities.

In different contexts and circumstances, such groups may have their roots in political ideologies, religious beliefs, ethnic, nationalist or cultural identities or, more likely, a complex mixture of several of these. They may focus on individuals or groups, but the most common feature is likely to be an opposition to existing centres of power.

They may be sub-state groups directed against the elites in their own state or foreign interests, or they may hold power in states in the south, and will then no doubt be labelled as 'rogue' states as they direct their responses to the north. What can be said is that, on present trends, anti-elite action will be a core feature of the next 30 years, driven in part by widening socioeconomic divisions exacerbated by environmental constraints.

Furthermore, such movements may, in some cases, engage in extreme action, including the willingness to cause mass casualties. They may well draw their leadership from well-educated and relatively wealthy sectors of society. If they succeed in effecting anti-elite revolutionary change, they may introduce rigid and authoritarian regimes in which they themselves constitute new elites. For the most part, the movements and resulting centres of political power will be male-dominated.

Such trends are in progress in the context of a demonstrated capability of political violence to affect centres of power, although some notable attempts of the mid-1990s did not achieve their full impact. Even so, there are many examples, from the Sendero bombing of elite neighbourhoods in Lima, through to the bombing of the Colombo central business district by the LTTE. An Algerian group made a failed attempt to crash a hijacked plane in the centre of Paris in 1994, and a Japanese religious sect attempted to kill thousands by releasing nerve gas into the Tokyo subway the following year. In the mid-1990s the Provisional IRA undertook a sustained programme of economic targeting of city-centre and transport network bombings in the UK, having a substantial if largely unacknowledged political impact.[5]

US forces overseas have experienced political violence over two decades, from the Beirut bombing of the embassy and marines barracks in 1983 through to the Khobar Towers and USS *Cole* bombings the following decade and the attacks on

[5]For an analysis of the use of economic targeting by paramilitary groups see Paul Rogers, 'Political Violence and Economic Pargeting – Aspects of the Provisional IRA Strategy, 1992–97', *Civil Wars*, vol. 3, no. 4 (2000), pp. 1–30.

embassies in Nairobi and Dar es Salaam. Perhaps most significant was the attempt to destroy the World Trade Center in 1993, an attack which, if it had succeeded, would have killed ten times as many people as the atrocities of September 11, 2001.

There is thus a strong record of anti-elite action by radical groups seeking political change, although this was not widely recognized prior to September 11. What is even less recognized is the connection between such action and the wider issue of the changing security parameters, even though such an analysis is not new. An early example was the work of Edwin Brooks, more than 25 years ago, who was concerned at the risk of

> a crowded, glowering plant of massive inequalities of wealth buttressed by stark force yet endlessly threatened by desperate people in the global ghettos…[6]

12.2 The US Security Paradigm

How does this analysis of international security trends compare with the outlook from the West in general and the United States in particular? Is there any indication that the core 'drivers' of potential insecurity need to be addressed, or is it a matter of maintaining the existing order?

President Clinton's first CIA Director, James Wolsey, characterized the post-Cold War security challenge for the United States by saying that the US has slain the dragon (the Soviet threat) but now lives in a jungle full of poisonous snakes.[7] The jungle includes the two potential rivals of Russia and China, at least in the long term, and two zones of continuing instability, north-east Asia and the Middle East. The latter is particularly significant because of the remarkable concentration of oil reserves. The less clear-cut threats include terrorism and disorder stemming both from 'rogue' states and failed states, especially where they affect US interests. This outlook would be accepted, to an extent, by principal US allies in Europe and elsewhere, but usually in a less clear-cut form and with certain concerns about the lack of a nuanced view, not least in relation to the Middle East.

In the US, the post-Cold War military posture changed radically during the 1990s. The army, navy and air force all lost around 30 per cent of their personnel while transforming themselves in the direction of rapid deployment capabilities, long-range force projection, counterinsurgency and special operations. The Marine Corps, with its amphibious capabilities, retained most of its strength. Military budgets may have been cut, but the resultant forces are considerable, especially in the absence of any kind of threat from the defunct Warsaw Pact, and are, by a very substantial measure, the most powerful in the world today.

---

[6]Edwin Brooks, 'The Implications of Ecological Limits to Growth in Perms of Expectations and Aspirations in Developed and Less Developed Countries', in Anthony Vann and Paul Rogers (eds) *Human Ecology and World Development* (London and New York: Plenum Press, 1976).

[7]Statement by James Wolsey at Senate Hearings, February 1993.

Some major NATO allies have followed a similar path, but the United States is pre-eminent, the overall direction being to sustain and enhance a capability to intervene in regional conflicts where national interests are considered to be at risk. Taming the jungle has replaced slaying the dragon. The exact nature and capabilities of those 'snakes' may not be too clear, a problem expressed with remarkable candour by President Bush in one of his early campaign speeches (January 2000):

> it was a dangerous world and we knew exactly who the 'they' were. It was us versus them and we knew exactly who 'them' was. Today we're not so sure who the 'they' are, but we know they're there.

In spite of the many military changes, the view on the Republican right in the latter part of Clinton's second term was that the US was not properly fulfilling its role of ensuring the enhancement of US interests internationally, and a markedly unilateralist stance developed initially in the Republican-controlled Congress. This is part of a wider view that the twenty-first century should be seen as the American Century.[8] Prior to the election of President Bush, this mood had been demonstrated by the refusal of the Senate to ratify the Comprehensive Test Ban Treaty, and by opposition to proposals for an anti-personnel landmine ban and moves towards an international criminal court.

After the election and prior to September 11, the unilateral stance included withdrawal from the Kyoto Protocols on climate change, a disinterest in Middle East peace proposals, a cessation of the diplomatic engagement with North Korea, opposition to UN proposals on the control of terrorism and an increasingly critical stance on the negotiations to strengthen the 1972 Biological and Toxin Weapons Convention. Not that the United States was unilateralist in all things, a point made admirably by Charles Krauthammer, writing in June 2001:

> Multipolarity, yes, when there is no alternative. But not when there is. Not when we have the unique imbalance of power that we enjoy today - and that has given the international system a stability and essential tranquility it had not known for at least a century.
>
> The international environment is far more likely to enjoy peace under a single hegemon. Moreover, we are not just any hegemon. We run a uniquely benign imperium.[9]

Thus, where it is in US interests to have agreements, then they are acceptable, as with NATO expansion and NAFTA. But the policy is highly selective and fits a paradigm in which US security interests are paramount and where the only way to ensure peace and prosperity is for the US to have freedom of action, whatever the effects on the world in general and even its allies in particular.

---

[8]See the Project on the New American Century, an interest group promoting a vigorous US foreign policy, with supporters including Vice President Cheney and Secretary of Defense Rumsfeld.

[9]Charles Krauthammer, 'The Bush Doctrine: ABM, Kyoto and the New American Unilateralism', *Weekly Standard*, June 4, 2001, Washington, DC.

12.3 The Implications of September 11

There was remarkable support for the United States in the immediate aftermath of the atrocities of September 11, 2001, together with an expectation that one result would be a recognition in Washington of the need for close multilateral cooperation in any response. In the event, the response, outside of intelligence and security collaboration, was of a substantial military operation that was primarily unilateral. This involved joining opposition forces in the Afghan civil war to defeat the Taliban regime, using a sustained bombing campaign that also resulted in the deaths of several thousand civilians. Most of the Taliban militia withdrew without surrendering and melted back into their communities, with very little of the leadership killed or captured.

This applied also to the al-Qaeda network which had its facilities in Afghanistan hugely disrupted, but lost few of its leaders and dispersed successfully to the extent that a senior FBI official concluded that it had been cut back in its potential for action by perhaps 30 per cent.[10]

In the immediate aftermath of the most intense phase of the war, Afghanistan experienced a degeneration into warlordism and banditry except in a few centres of population, notably Kabul, where an international security assistance force provided some stability. Meanwhile, the United States was able to extend its bases into Afghanistan, Tajikistan, Kyrgyzstan and Uzbekistan, to maintain a military presence in Pakistan and even to obtain basing facilities for tanker aircraft in Bulgaria, the first presence of foreign troops in that country since 1946.

More generally, the September 11 attacks and the resulting war on terror have greatly reinforced the unilateralist outlook. In the wider context, this has included strong opposition to further multilateral negotiations on strengthening the bioweapons convention, the establishment of military tribunals meeting in secret to try presumed terrorists and the withdrawal from the Anti-Ballistic Missile Treaty. This last decision, in early December 2001, coincided with a large 'ripple-fire' test of the Trident strategic nuclear missile system, a potent reminder of the US desire to acquire missile defences while retaining the world's most powerful offensive nuclear forces.

In relation to the 'war on terror' itself, there has been an intensification in both words and action. As well as the establishment of an extended network of bases across Central Asia, increased military assistance has been given to the Philippines in its counter-insurgency operations against Islamic militants. The Fiscal Year 2003 defence budget will see a very substantial increase in military spending in addition to increases already agreed for 2002. The 2003 budget will rise by a figure nearly 150 per cent of that of the entire UK defence budget, with considerable emphasis on those aspects most relevant to low intensity and counter-insurgency operations. Maintaining control has become the dominant theme of the security paradigm.

---

[10]The view of the acting assistant director of the FBI's counter-terrorism division. See Walter Pincus, 'Al Qaeda to Survive bin Laden', *Washington Post*, December 19, 2001.

The rhetoric, too, reflects these budgetary developments. The underlying theme is more than the well-rehearsed 'if you are not with us you are against us', for there is a clear emphasis on Washington deciding who is 'against us'. Most notable was the 2002 State of the Union Address in which President Bush referred to an 'axis of evil' encompassing North Korea, Iraq and Iran; the latter being included greatly to the surprise, and consternation, of European diplomats.

12.4 Future Trends in Political Violence - and Alternatives

What is the relevance of the attacks of September 11 and their after-math to the more general analysis of trends in international insecurity? Will it make a re-examination of the Western security paradigm more or less likely? The first point to emphasize is that the al-Qaeda attacks did not come from a desperate underclass of marginalized peoples across the globe, driven to violence by frustration of the possibilities of peaceful change. Even so, the context of the development of the network is relevant in several respects.

First, the September 11 attacks did demonstrate in a horrifying manner the vulnerability of highly developed states to asymmetric paramilitary action. Second, al-Qaeda's support stems, in part, from an opposition to the US military presence in the Gulf, boosted by disaffection from among the 'demographic bulge' of educated young people in the region. Finally, and most significant, the reaction of the United States has not been to address the core issues underlying the strength of this and other networks, but rather to seek to maintain control with great vigour and expanded military force.

The more general analysis suggests that the core drivers of socioeconomic divisions and environmental constraints will lead to increased instability and conflict, much of it represented by an increasing tendency towards anti-elite insurgencies. These will, on occasions, result in considerable violence, including the risk of mass casualty attacks similar to or worse than those of September 11. In turn, there will be vigorous counter-insurgency actions, both by the military and paramilitary forces of local elites, and by substantial action by the US in its war on terror. The longer-term result is likely to be a cycle, or perhaps a spiral, of violence between elites and radical anti-elite movements.

While a short-term alternative might involve far greater cooperation for anti-terrorist actions with a strong emphasis on multilateral approaches, longer-term action requires some fundamental changes in Western policies aimed at reversing socioeconomic divisions and engaging in effective environmental management linked to sustainable economies. This requires radical changes to policies on the debt crisis, the promotion of trade reforms specifically linked to the trading prospects of the South and the encouragement of economic cooperation for gendered and sustainable development. It also requires a radical reappraisal of northern attitudes to economic growth, centred on effective action on climate change and excess resource use.

In the aftermath of September 11, the short-term prognosis for such a change in the paradigm is negative. In the United States, at least, the combination of a notably right-wing administration, especially in security terms, with the traumatic and far-reaching domestic effects of the attacks, has produced a reinforcement of the existing paradigm of maintaining control, keeping the lid on problems of instability and threats from the jungle full of snakes. There is little or no attempt to understand or address the core issues.

12.5 Endless War?

While sympathy for the United States over the effects of the New York and Washington attacks remained strong in Europe for some months, there was increasing concern over the unilateralist stance, both before and after September 11. To varying extents, European governments have taken a different view from that of the US on a wide range of issues: from arms control issues such as missile defence, the test ban treaty and the control of bioweapons, anti-personnel landmines and light arms transfers, through to a greater concern over stability in Northeast Asia and the Middle East. There remains a broadly shared view on some global issues such as WTO, IMF and World Bank activities, but European concern is much higher on issues such as debt and climate change. All this is within the context of a real concern over the clear determination of the US to act as a single-minded superpower, with little interest espoused for the views of its putative allies.

More widely, there is in the majority of the world a developing analysis that is wildly different from the world view from Washington. One analysis, published soon after the attacks, condemns them as horrific, despicable and unpardonable, but cautions against an automatic 'iron fist' response that ignores the underlying context. It points to the frequent use of indiscriminate force by the US and to the bitter mood throughout much of the Middle East and Southwest Asia, directed partly at the US but also against autocratic states dependent on continuing US support. The analysis concludes:

> The only response that will really contribute to global security and peace is for Washington to address not the symptoms but the roots of terrorism. It is for the United States to re-examine and substantially change its policies in the Middle East and the Third World, supporting for a change arrangements that will not stand in the way of the achievement of equity, justice and genuine national sovereignty for currently marginalised peoples. Any other way leads to endless war.[11]

The prognosis, at least for the present, has to be one of endless war, with a substantially increased risk of political violence, both by elites and against elites. An alternative possibility is that the trauma of September 11 might encourage

---

[11]Walden Bello, 'Endless War', *Focus on the Global South*, Manila, September 2001. This and other analyses are available at www.focusweb.org.

individuals, citizen groups, intellectuals and indeed political leaders to recognize the long-term security significance of what has happened and to redouble efforts towards a more equitable and stable world.

It is essentially a matter of choice, and the next decade is likely to prove pivotal in determining the degree of international instability and political violence. The early effects of September 11 suggest a hardening of the old paradigm, but there is every chance that it may be possible to further analyse and demonstrate the futility of that approach. The responsibility for those in a position to do so, whether academics, activists, politicians or many others, is considerable.

Chapter 13
Iraq: Consequences of a War (2002)

The commitment of the Bush administration to regime termination in Iraq is so strong that a war with Iraq early in 2003 is now highly likely. Regime termination is thought necessary, in part, because of Iraq's likely development of weapons of mass destruction, but the control of the region's oil resources is also a major factor.[1]

Although the regime is weaker than eleven years ago, it is probable that most of its elite forces will resist attack. The regime will seek to draw foreign forces into a war in the greater Baghdad region. US war plans are expected to focus on destroying Iraq's military and administrative infrastructure, denying the regime access to its energy supplies and drawing out the elite forces so that they can be subject to intensive air attack.

Conflict in Baghdad will involve the use of area-impact munitions as well as precision-guided munitions and the civilian casualties will be high. A civilian death toll of at least 10,000 should be expected but this may be a low estimate, given the experience of urban warfare in Beirut and elsewhere.

Evidence from the 1991 war indicates that it is highly likely that the regime will use all available military means, including chemical and biological weapons, if its very survival is threatened. Such weapons may be used in tactical warfare to hinder invading forces but may, in extreme circumstances, be used strategically against forces in other countries.

In such circumstances, and especially if high casualties result, there is a possibility of a nuclear response. The British Government, in particular, has been candid in conceding this as a potential response to CBW use by Iraq.

The United States has sufficient forces to ensure regime destruction but the regime's replacement by occupying forces or by a client regime, even if the war is

[1]This text was first published as: "Iraq: Consequences of a War", *Briefing Paper*, Oxford Research Group, October 2002 of which the author retained the copyright. The author would like to thank General Sir Hugh Beach, Dr. Scilla Elworthy, Air Marshal Sir Timothy Garden, Professor Bill Kincaid, Susan Kincaid and Dr. Jack Mendelsohn for commenting on a draft of this report, and Nick Ritchie for his help with the production. He takes take full responsibility for its contents.

P. Rogers, *Paul Rogers: A Pioneer in Critical Security Analysis and Public Engagement*, Pioneers in Arts, Humanities, Science, Engineering, Practice 21,
https://doi.org/10.1007/978-3-319-95150-8_13

not greatly destructive, should be expected to increase regional opposition to the US presence. It is likely, in particular, to increase support for organisations such as al-Qaida and to prove counter-productive to peace and stability in the region.

Given these major consequences of a war – high civilian casualties, risk of the use of weapons of mass destruction and post-war regional instability – alternatives to war should be sought as a matter of urgency.

13.1 Introduction

While there are still indications of differences between some of the senior US military and the civilian security establishment in the Bush administration, it is clear that it is the latter who are dominant in the argument over how to handle the Saddam Hussein regime. For them, it is not sufficient for the UN inspectors to be allowed back into Iraq, even without restrictions – they believe US policy should be clearly to terminate the regime.

There is thus a very firm determination in Washington to change the leadership in Iraq, replacing it with military occupation or a pro-western regime. Indeed, it is even believed in some circles that the removal of Saddam Hussein would actually improve the prospects for a peace process between the Israelis and the Palestinians.

The timing of that intervention or forced regime change remains unclear, and is dependent to an extent on attitudes and resolutions at the United Nations, but substantial military action is most likely in January or February 2003 if Saddam does not accept unfettered and coercive inspections.

In the face of a potential new Middle East war, there are a number of key questions to ask.

Why is the United States determined to destroy a regime that has already been weakened by eleven years of sanctions, what form would the attack take, and how might the regime respond to its imminent destruction?

A reasonable prognosis for the conflict can be developed based on the war aims of both parties and their levels of military capability. This can be further aided by examining relevant elements of the 1991 war, how the Iraqi regime acted in relation to the UN inspection process during the 1990s, and how other conflicts have been fought, not least elsewhere in the Middle East and South Asia.

This briefing addresses these issues but its principle aim is to make an assessment of the possible consequences of a war in terms of three issues – civilian casualties, the risk of escalation to the use of weapons of mass destruction and the state of the region should the regime be terminated and replaced with a regime acceptable to Washington.

The briefing seeks to avoid what are often termed “worst case scenarios”. Some of these are discussed briefly in one section of the paper, but it aims primarily to concentrate on probable rather than possible outcomes.

13.2 Why Attack Iraq Now?

It is first relevant to examine the reasoning behind the intended US action. Clearly it is not just about the UN inspection process, for there is a palpable sense in Washington that full access for the inspectors is just about the last thing that is wanted. Part of the reason for this is that the security community in the Bush administration does not believe that inspections can uncover everything and that the Iraqi regime is so determined to develop and deploy weapons of mass destruction that it will eventually hinder the inspection process.

In this respect, timing is crucial. Military action against Iraq intended to destroy the regime could take up to three months and needs to commence before the end of next February in order to avoid combat under markedly adverse climatic conditions. If the war is not started by then, the US would have to consider a delay until the autumn, and this presents three serious problems.

One is maintaining substantial forces in the region throughout the summer months and a second is being engaged in such a war as the 2004 Presidential elections get under way.

Perhaps the most substantial problem is that President Bush and his associates have been unequivocal in the need to deal with the Saddam Hussein regime. Any substantial delay would look like political ineptitude by the Bush administration and would allow time for domestic opposition to develop.

There are other issues that go well beyond the timing of an attack, and some relate to the experience of the 1991 war, a war that appears to have ended in a complete victory for the coalition but is viewed in a rather different light in security circles. There remain powerful memories of that war in that it was known that the Iraqis had chemical and biological weapons and were ready to use them if the regime was threatened with destruction. The view by the hard-liners in the Bush administration is that the Iraqis either have such weapons now or will have them very soon, so early pre-emption is essential.

Those memories of 1991 are also relevant in two other ways. One is that what appeared to be a total victory was achieved partly because the Iraqis did not even commit their best troops to the war zone, regime survival being their primary goal. Those troops were then available to destroy the Shi'ite and Kurdish rebellions.

A second aspect is that the missiles fired against Israel and Saudi Arabia in 1991 were a damaging diversion for the US. It is recognised, in particular, that the missile that just missed an ammunition depot at Jubayl during the closing stages of the war could have caused a catastrophe.[2] The Iraqi policy of withholding its best troops in

---

[2]On 16 February, 1991, a missile narrowly missed a large pier complex in the Saudi port of al Jubayl. It landed in the sea some 300 m from the US Navy's aviation support ship *Wright* and close to the large amphibious warship *Tarawa*, both of which were moored alongside a pier complex which included a large ammunition storage dump and a petrol tanker parking area. Details of this incident only entered the public domain several years later but it is credited with having a considerable impact on US military attitudes to states armed with ballistic missiles. See:

1991 supports the idea that its military have a fair degree of tactical competence and could therefore be a threat in the future. This, along with the effect of the Scuds, and Iraq's possession in 1991 of CBW systems, does appear to demonstrate the capacity of relatively weak states to deter even a superpower, an entirely unacceptable situation for the United States.

The Bush administration is particularly firm when it comes to international security, and this has been much strengthened by the 9/11 attacks and the consequent and crucial need to regain control of a volatile international system after such a shock. The unilateralist outlook places US security interests at the forefront of foreign policy[3] and it is simply not acceptable to have rogue states in an axis of evil that are able to develop weapons of mass destruction that might deter the US from necessary interventions. It is implicit in the new security posture that such circumstances may need to be dealt with by pre-emption. Moreover, in this respect, terminating the Saddam Hussein regime would set a very firm example to other regimes, not least Iran. As one influential Republican with impeccable oil links put it: "If we sort out Iraq right we may not even have to go for Iran".[4] A further factor is the political utility of "going to war". It demonstrates taking the battle to the enemy, mobilises the population and distracts the public from domestic worries.

13.3 The Relevance of Oil

Yet there is a further reason that receives far less attention – the regional geopolitics of oil. As an earlier Oxford Research Group briefing paper emphasised,[5] the Persian Gulf oil reserves are truly staggering – twice as large as all of the rest of the world's oil fields put together, and more than twenty times larger than the dwindling

John D. Gresham, "Navy Area Ballistic Missile Defence Coming on Fast", *Proceedings of the United States Naval Institute*, January 1999.

[3]The development of the current US security paradigm is outlined in an earlier paper in this series: Paul Rogers and Scilla Elworthy, *The United States, Europe and the Majority World after 11 September*, Oxford Research Group Briefing Paper, October 2001.

[4]Interview with member of the Bush transition team, Washington DC, November 2001.

[5]Paul Rogers and Scilla Elworthy, *A Never-Ending War? Consequences of 11 September*, Oxford Research Group Briefing Paper, March 2002. As well as drawing on the experience of US operations in the first Gulf War, in Serbia and in Afghanistan, this section is compiled from two principal sources. The first is a wide range of reports from *The Washington Post* and the *New York Times* and agency reports from Reuters and Agence France Presse, most of these being available at www.globalsecurity.org. Secondly, it uses material from a range of defence publications, particularly *Defence News*, *Jane's Defence Weekly* and *Aviation Week and Space Technology*. Three articles in the last of these were of particular value: David Fulghum, "Iraq Strike Has Focus, But There's No Timetable (5 August, 2002); David A. Fulghum, "War Planning for Iraq Continues on Target" (23 September 2002); David Fulghum and Robert Wall, "Oil Wells, WMD Sites Fall Off Bombing List" (30 September 2002).

reserves of the United States itself. Iraq alone has close to four times the oil reserves of the United States including Alaska.

Among the Gulf States, Saudi Arabia is by far the most important, yet there is increasing evidence of internal instability, a deep undercurrent of anti-American sentiment, and even evidence that much of the financial and other support for al-Qaida comes from within the Kingdom. Second only to Saudi Arabia as a repository of easily extracted high quality oil reserves is Iraq. Replacing the current regime with one compliant with US interests would therefore involve the substantial bonus of diminishing the importance of Saudi Arabia. This could be seen, perhaps, as the real al-Qaida link with the Iraq question.

The significance of oil supplies is put in further perspective when we examine trends in US import dependency. Until just over thirty years ago, the United States was able to satisfy its substantial oil requirements from its large domestic reserves, but production has tended to decrease throughout the past thirty years with consumption increasing. The end result has been a rapid increase in oil import dependency, with the current import requirement being in excess of 60 per cent of demand. Furthermore, there is little inclination to curtail domestic oil consumption, especially under the Bush administration, and even the North East Alaska fields would add relatively little in the face of the high level of domestic demand.

All of this must be put in the context of world oil trends where two features are relevant. The first is that the most promising area for oil prospecting is the Persian Gulf region, and the oil that appears available there is of a relatively high quality, is easily recoverable so that production costs are low and is close to major sea-lanes. The known reserves of oil in Saudi Arabia and Iraq are both greater than they were a decade ago in spite of substantial annual production from both states – the discovery of new reserves is actually exceeding production, in marked contrast to the situation in the United States.

Secondly, new reserves being discovered and developed elsewhere in the world are most commonly expensive to extract or are of relatively low quality. Alaskan, North Sea and West Shetland oil all require expensive extraction technologies operating in adverse conditions, Caspian Basin oil requires long distance transport to sea routes and Mexican and Venezuelan oil is of relatively lower quality.

In any case, other supplies need to be put in perspective. If one adds together the entire proved oil reserves of the Caspian Basin, Siberia, Alaska, the North Sea and West Shetland, one gets a combined total that is rather less than that of Iraq alone, and Iraq, while very oil-rich, has less than one-fifth of the total reserves of the Persian Gulf region. When we combine US unilateralism with a specific unwillingness to allow opponents to develop deterrent forces and then add in the importance of Iraqi oil, we get some idea of the reasons for the determination of the administration to go to war with Iraq. Within this outlook there lies the belief that such a war will be winnable and that a US ally in Iraq will ensure that other states in the region will acquiesce to US interests in the face of this determination. Thus the region will be made safe for the West in general and the United States in particular.

Termination of the Saddam Hussein regime is therefore central to America's evolving policy of maintaining control of a volatile region that is crucial to international energy supplies. It is also an essential part of controlling the evolution of oppositional states that might develop the means to deter the United States from taking actions considered essential to its security.

13.4 What Kind of War?

In preparation for the war, the US has accelerated production of precision-guided weapons, partly to replace those used in Afghanistan but mainly to have large enough stocks for use against Iraq. Munitions plants are currently working 24 hours a day, seven days a week, and production of protective chemical and biological warfare suits has also been increased.

Supplies are already being moved into the region, a large air base in Qatar is being prepared as an alternative to Saudi facilities and the existing aircraft carrier battle group in the Gulf is likely to be augmented by up to three more by early next year before a war starts, most probably in January or February.

Many army and marine corps units are currently training for desert and urban operations, and there is a particular emphasis on special forces and the equipment needed to move them around within Iraq. In a move unreported in the mainstream media, the United States already has repair teams actually inside Iraq, working on three airfields that are in areas in the North that are controlled by the Kurds. Furthermore, the intensity of US and UK air attacks in the no-fly zones has increased substantially.

The exact nature of the attack is beginning to become clear and looks like being a combination of the different options that have been discussed so widely in the press. A core component will be the massive use of air power to destroy the command and control systems of the Iraqi military, using cruise missiles, stealth bombers and strike aircraft and B-1B and B-52 strategic bombers. The latter will operate from bases that include the British territory of Diego Garcia in the Indian Ocean.

The US Air Force is currently building specialised facilities there to enable it to operate B-2 stealth bombers that currently operate only from a single base in Missouri. The B-2 would be a key component of an air assault but it has a low level of reliability, with barely a third of the force of 21 planes available for combat operations at any one time. Because of this, and the need to operate over very long distances, the B-2 was only able to be used sparingly in Afghanistan. Forward basing to Diego Garcia is a response to this predicament.

During the air war itself, air defences, air bases, army barracks and presidential palaces will all be hit, and power supplies throughout much of the country will be destroyed. Some of the cruise missiles may disperse long carbon fibre strands to short-circuit electrical switching stations, and a form of bomb first used in Serbia will have a similar function.[6]

The experience of the 1991 Gulf War, combined with US tactics used in Serbia, indicates that the air war will extend its targeting to a very wide range of Iraqi state functions that could contribute in any significant way to the survival of the regime. Among these will be any manufacturing facility that has a defence connection, whether this is direct in the form of armaments, or indirect in terms of component manufacture. There is also likely to be a sustained programme of the dismantling of those aspects of the state concerned with national organisation and control, including a wide range of government ministries. Especially significant will be any targets connected with the central security apparatus, including the Republican Guard, the Special Republican Guard and the five different security and intelligence organisations (see below).

Given that this war, unlike the 1991 war, has the specific aim of destroying the regime, the air war against its core infrastructure is likely to be substantially more intense than before. It will also use a range of new technologies.

A new type of weapon, a version of the Wind Corrected Munitions Dispenser (WCMD) will be used that disperses huge numbers of microscopic carbon fibres that drift in the wind. These can get into even the smallest of electrical components, damaging computers, air conditioners, communications equipment and anything else with electrical circuits.[7]

Another new type of weapon will be used extensively, the high-power microwave weapon that produces a near-instantaneous electrical pulse or 'spike' that destroys computer memories and damages electrical components. Microwave weapons will be used, in particular, against sites believed to contain chemical or biological weapons in an effort to prevent these being used.

13.5 Area Impact Munitions

It is appropriate to emphasise the fact that area-impact munitions[8] are expected to play a substantial role in the war, whereas the impression is frequently given that munitions will be limited almost entirely to precision-guided missiles and bombs aimed precisely at point targets.

---

[6]"Wind-corrected munitions kit goes operational on F-16 and B-52", *Flight International*, 14 November 2000.

[7]"Washington Outlook", *Aviation Week and Space Technology*, 2 September 2002; David A. Fulghum, "USAF Acknowledges Beam Weapon Readiness", Aviation Week and Space Technology, 7 October 2002.

[8]An account of the development and deployment of area impact munitions after the Vietnam War is contained in a chapter on conventional weapons of mass destruction in: Paul Rogers and Malcolm Dando, *The Directory of Nuclear, Biological and Chemical Arms and Disarmament, 1990*, Tri-Service Press, London, 1990.

In practice, the remarkable improvements in munitions accuracy over the past twenty years has been paralleled by the development of area-impact munitions (AIMS), the modern-day successors to the napalm used in Korea, Vietnam, Afghanistan and elsewhere. AIMS are designed specifically to damage and destroy "soft targets" (including people) and take three main forms, cluster bombs, fuel-air explosives (also termed thermobaric weapons) and multiple-launch rocket systems – all were used in the 1991 war and, more recently, in Afghanistan.[9]

A standard British cluster bomb, the BL755, disperses 147 bomblets over an area of several thousand square metres, with each bomblet producing up to 2,000 high-velocity shrapnel fragments. A US Air Force fuel-air explosive is the CBU-55B cluster unit containing three BLU-73B canisters, each with over 30 kg of ethylene oxide. Each canister produces an aerosol cloud about 20 m in diameter and metres thick which is detonated to produce blast overpressures of up to 300 psi.[10]

An even more powerful weapon is the BLU-82B ("Big Blue") slurry bomb originally developed for clearing Vietnamese rain forest terrain to create 'instant' helicopter landing zones. This weapon, the world's most powerful conventional explosive, is a 6.8 tonne bomb containing ammonium nitrate, powdered aluminium and a polystyrene soap binding agent, all in aqueous solution. When detonated, the BLU-82B produces blast overpressures of up to 1,000 psi, close to those of a small tactical nuclear weapon.

The BLU-82B was used against Iraqi bunkers in the 1991 war and was used extensively in Afghanistan against cave complexes. It is likely to be used in Iraq in the event of a war, especially against hardened targets. If these are in urban areas, as is likely, the collateral damage among the civilian population may be very high.

The US MLRS is an example of a rocket launch system. It comprises air-mobile launchers that would be used in a ground war, with each launcher capable of ripple-firing a salvo of twelve missiles over a 30+ kilometre range, dispersing nearly 8,000 sub-munitions over an area of 20 ha (nearly 45 acres), with each sub-munition dispersing large numbers of high-velocity shrapnel fragments. If these were to be used against urban targets, once again, the collateral damage would be likely to be substantial.

13.6 A Ground War

The air attacks are confidently expected to do massive damage to the regime's ability to use its security forces, especially the Republican Guard, but it is recognised that this will not, in itself, destroy the regime. For that to happen, there has to

---

[9]Further thermobaric weapons are currently under development and may be ready for deployment in a war against Iraq. See: Andrew Koch, "USA speeds development of thermobaric weapons", *Jane's Defence Weekly*, 28 August 2002.

[10]David A. Fulghum, "Iraq Strike Has Focus But There's No Timetable", *Aviation Week and Space Technology*, 5 August 2002.

be substantial US intervention on the ground, although this is unlikely to involve a major conventional assault on Baghdad until well after the commencement of the air war.

It is probable that there will be three forms of ground action.[11] One will be an intervention in the south-eastern region of Iraq, including Basra, to ensure control of the major oil fields there. This may follow the disabling of the oil production facilities by air attack, with an emphasis on the use of disabling weapons rather than the actual destruction of the facilities. A second will be a combination of air attacks together with sufficient troops to control the oil-producing regions of the Kurdish north, protect the newly established air bases and prevent any kind of Iraqi counter attack.

Operating strike aircraft and attack helicopters from within Northern Iraq is intended as a militarily useful operation but is also seen as having a strong psychological impact on the supporters of the regime.

Both of the actions, in the North-West and the South-East, are expected to be opposed, especially that around Basra, and there will be very heavy use of air strikes to limit US casualties. High levels of co-lateral damage and civilian deaths are almost inevitable, not least in and around Basra where urban conflict is possible.

The third ground action will be the use of highly mobile troops in moves towards Baghdad itself. Their primary aim will not be to destroy the regime directly, but rather to force it to bring the Republican Guard Divisions and Special Republican Guard Brigades out into the open. There, they will be exposed to the full force of US air attack, some of it operating out of Northern Iraq itself. As well as the use of precision-guided munitions, this will include the extensive use of carpet bombing with conventional bombs, as in Afghanistan, as well as area-impact munitions such as cluster bombs and fuel-air explosives, as outlined above.

One significant point is that most of the ordinary Iraqi army units will not be subject to attack unless they appear a threat to US troops.[12] The thinking in Washington is that they will be required to help stabilise Iraq once the regime has fallen.

Within a matter of days, it is believed that Iraqi military communications will be virtually defunct, the regime will be cut off from its oil supplies and the elite Republican Guard forces will be in the process of being destroyed.

Within weeks, and certainly before the summer, the regime will be finished, and will be replaced by a US occupying force, possibly using some Iraqi army elements. Acceptable leadership may eventually emerge, drawing on elements of the regular army to establish post-war stability, in a client relationship with Washington.

It hardly needs to be said that the side effects on the Iraqi economy and the well-being of its people will be devastating. One of the core US requirements will

---

[11]Ibid.

[12]A detailed assessment of Iraq's security and intelligence forces is in: Ibrahim Al-Marashi, "Iraq's Security and Intelligence Network: A Guide and Analysis", *Middle East Review of International Affairs*, Vol. 6, No. 3, September 2002.

be to minimise the casualties among its own troops and this will involve the massive and continual use of air power. Experience of the tactics used in the first Gulf War and in Serbia shows that, quite apart from large numbers of civilians killed and injured directly in the attacks if urban areas are attacked with area-impact munitions, there will certainly be substantial and sustained damage to the economy as power supplies, transport and communications are destroyed.

Without electrical power, sewage and water purification plants will not operate. As bridges and other transport nodes are destroyed, the domestic economy will grind to a halt. As in 1991, the effects will be felt by ordinary people for months and years afterwards unless there is a substantial international post-war effort to revive and restructure the nation and its economy.

Even so, there is a belief that the war is a price worth paying, and that the end result will be a pliant regime, sympathetic to and dependent on Washington for security, and controlling some of the world's richest oil reserves. The US position in the Middle East will be secure and an effective example will have been made of an unacceptable threat and an unacceptable effort to develop weapons of mass destruction. Perhaps most important of all, there is a belief among the Bush administration's security advisers that other states will have been impressed by American determination and will be far more cautious in their approach to regional security.

13.7 The War from Baghdad

What of the view from Baghdad? Almost certainly, the dominant strand of thinking within the Saddam Hussein regime is the imperative for regime survival. This must not be underestimated – it transcends every other objective and was evident in 1991 when the conscript defenders in and around Kuwait were sacrificed to preserve the regime.

In recent years, Baghdad has worked diligently to improve its relations with neighbouring states. This has not always been easy, especially with Iran, but the overall effect has been to ease tensions, not least by showing evidence of support for the Palestinians that has gone down well in terms of Arab public opinion. Moreover, it has made Saudi Arabia deeply reluctant to become involved in support of the United States.

Whatever else happens, their support will be at a much lower level than in 1991.

The regime has also sought to encourage opposition to the war from France and Russia, not least by offering contracts and possibly by indicating that past debts may be repaid. It has also sought Chinese aid in rebuilding its air defences, conscious of China's long-term interest in Gulf oil as its own reserves run down.

The Saddam Hussein regime has been helped in its aim to represent an attack on Iraq as an attack on the Arab world by the widely perceived support that Washington has given to the Sharon government in what is seen as a particularly hard-line policy towards the Palestinians. The fact that it is American helicopters and strike aircraft that the Israelis use persistently in attacks on Palestinian targets lends support to this view.

Iraq will also prevaricate and procrastinate mightily on the matter of inspections, continuing to make a range of offers that attract international approval. Its aim will be to postpone a military confrontation until the autumn of next year, for reasons outlined earlier.

There remains controversy over whether the regime really does have chemical and biological weapons (CBW), but it must be remembered that the UNSCOM inspectors only really ran into trouble in the late 1990s when they appeared to be getting close to these weapons and their production facilities. Iraq had CBW in 1990–91 and both the UK government and the CIA say they have available CBW now.

From the regime's point of view, CBW systems constituted a last ditch deterrent in 1991 and could do so again, although there would also be seen to be a military utility to their use, especially with fast-acting chemical warfare agents. The US military expect CBW use and are planning for it. If they are right that the regime has useable CBW, then full inspections by UN staff will not be allowed by the regime, even if prevarication lasts for some months.

The regime recognises that it cannot rely on the Iraqi people as a whole, only those who constitute the elite and have a vested interest in regime survival. The Iraqi army's strength is estimated at around 375,000 including recalled reserves, but the great majority of these are ill equipped and of poor combat efficiency. Six Republican Guard divisions (three armoured, one mechanised and two infantry) numbering perhaps 60,000 troops are better equipped but may be of questionable reliability. As such, they may not be used in the core defence of the regime.

More significant are four brigades of the Special Republican Guard, five brigades of commandos and two special forces brigades, together totalling perhaps 50,000 troops. In addition, the regime maintains five different security and intelligence organisations. The general security service numbers 8,000 and works principally to counter dissent in the general population. The general intelligence service is of a similar size and concentrates on internal and external intelligence. Military intelligence, rather smaller in number performs, a more military function, and there is a separate military security service that appears to have its own brigade. Finally, there is the special security force, working with the Special Republican Guard and numbering perhaps 5,000 people drawn heavily from Saddam Hussein's own district of Tikrit, north of Baghdad.[13] These security and intelligence forces together number perhaps 30,000 people, giving a total force of reasonable reliability of around 80,000.

The regime will be broadly aware of US plans for military action and will have undertaken a full-scale process of dispersal and disguise for its elite forces. It may well have back-up communications systems hardened against precision strike and microwave attack, not least through the recently installed fibre optic underground

---

[13]See, for example: Michael R. Gordon, "With his army weak, Saddam might force war into the cities", *International Herald Tribune*, 27 August 2002, and: Rajiv Chandrasekaran, "Iraqis would use urban warfare to trap U.S. troops", *International Herald Tribune*, 28 September 2002.

links, and may have dispersed its CBW systems to prevent their early destruction. It may also have pre-delegated authority to use these systems to regional or local commanders. This threat is recognised in Washington and London to the extent that the US has made it clear that any commanding officer using WMD will be judged as a war criminal after the conflict.

Iraqi tactics in response to a US attack will probably have several elements. One will be to make the war as difficult and as long-lasting as possible, even if this means allowing US troops into Baghdad itself and only engaging them at close range.[14] Such an engagement may well involve CBW weapons, but there may also be delivery systems available to direct these weapons at US bases in northern Iraq and in Kuwait and Qatar, perhaps pre-emptively, and possibly to engage Israel in the conflict.

The focus, though, will be on Baghdad, a sprawling city of nearly five million people. There will be a ready recognition that the elite armed forces and security units are the key to the conflict and that their survival will be hastened by dispersing them and drawing the US forces into urban combat. There may be limited use of chemical weapons early in this conflict, to force the US troops to fight in their restrictive CBW suits.

The extent of the fighting is almost impossible to assess, and depends partly on whether the majority of the Iraqi population supports the regime, or acts against it. The former may seem unlikely, but it is not easy to assess the ordinary people's response to what would clearly be a foreign force seeking to occupy the city. In any case, one possible effect of a war in Baghdad could be immediate and massive movements of refugees, complicating US war efforts.

What is important is that any major urban resistance would be countered by US firepower and would most likely cause many casualties. In this connection, the experience of the Israelis in West Beirut in 1982 is relevant. There, powerful and well-equipped Israeli armed forces, with complete control of the air, laid siege to a densely populated urban area defended by just 15,000 militia, mostly lightly armed. The Israeli armed forces anticipated that they could take the city within a week, but found it remarkably difficult to do so, even with the use of very heavy aerial and artillery bombardment. The siege lasted three months, some 20,000 people were killed, the great majority civilians, and the international outcry against the Israeli action was such that the PLO fighters were withdrawn under safe conduct.[15]

While the regime will seek in every way to extend the war, there will still be a quiet recognition that the United States does have the military means to destroy Iraq. Its forces and its economy are both weak after 11 years of sanctions, whereas the United States has even greater military strength. It follows that more vigorous

[14]Peter Sluglett and Marion Farouk Sluglett (editors), *The Times Guide to the Middle East*, Times Books, London, 1991.

[15]Paul Rogers, "Security Consequences of the Osiraq Raid", *Contemporary Security Policy*, Vol. 19, No. 2, 1998.

responses should be expected, bearing in mind the absolute requirement of regime survival. If it is clearly and unequivocally threatened, then almost anything goes.

One of the less-remembered aspects of the Gulf War was the firing of 360 Kuwaiti oil wells and the consequent destruction of five million barrels of oil and widespread environmental damage. A US invasion of the Iraqi oil fields around Basra might result in a similar sabotaging of these fields. If this was accompanied by paramilitary actions against Kuwaiti and Saudi refineries and oil fields, the effect on world oil prices would be considerable, especially if the targets could be contaminated with chemical or radiological weapons. Such contamination could be long-lasting, as could the effects on oil markets.

13.8 Chemical and Biological Weapons

According to the CIA and the British government's dossier on Iraqi weapons, Iraq currently has useable chemical and biological weapons, and, in the British view, a command and control system that probably involves delegation from the command authority. How such systems would be used will depend on the course of the war, but there is much to be learnt that is relevant from an understanding of the Iraqi development of such weapons in the period leading to the 1991 war.

Following the Israeli destruction of Iraq's Osiraq nuclear research reactor in 1981, the Iraqi regime engaged in a multifaceted and dispersed programme of producing nuclear, chemical and biological weapons. Particular emphasis was placed on biological agent developments in the late 1980s, and weaponisation of several agents, and of nerve agents, was underway towards the end of the decade.

Between August 1990, when Iraq invaded Kuwait, and the onset of Desert Storm in January 1991, the regime instigated the emergency completion of a limited capability for medium range offensive action with CBW agents. This resulted in the deployment to four remote locations in Iraq of a force of missiles and bombs. 166 bombs were completed, 100 filled with botulinum toxin, 50 with anthrax and 16 with aflatoxin. 25 Al Hussein missiles were deployed, 13 armed with warheads containing botulinum, 10 with anthrax and two with aflatoxin.[16]

These, and chemical warheads, were intended for use in the event of the destruction of Baghdad by nuclear weapons during the Gulf War. This could have resulted from an unforeseen escalation of the war, possibly involving Israeli action in response to Scud missile attacks. Moreover, authority to launch the missiles was pre-delegated from Baghdad to regional commanders, an extraordinarily risky

[16]Most of the information on CBW development and deployment came into the public domain four years after the Gulf War with the publication of a report from the UN Special Commission on Iraq: *Report of the Secretary-General on the status of the implementation of the Special Commission's plan for ongoing monitoring and verification of Iraq's compliance with relevant parts of Section C of Security Council resolution 687 (1991)*, New York, UN Security Council report S/1995/864, 11 October, 1995.

decision, giving rise to the possibility of unauthorised use of weapons of mass destruction during a fast moving and, for the Iraqis, chaotic and increasingly disastrous military confrontation.

Perhaps the most significant aspect of this deployment is that the US intelligence community was aware of the major features of the Iraqi BW programme, including the assessment that the Iraqis were likely to use weapons of mass destruction if the survival of the regime was threatened. This was the subject of a National Intelligence Estimate prepared in November 1990, three months after the invasion of Kuwait and two months before the onset of Desert Storm.[17]

It is relevant that coalition forces did not seek to destroy the Iraqi regime in March 1991.

While motives for this restraint included risk of coalition casualties and concern over the political dismemberment of Iraq, there are unofficial indications that the fear of an escalation to WMD use was also of concern.

UNSCOM investigations, especially during the period 1995–1998, established the extent of the remarkably large Iraqi BW programme, and it would be foolish to exclude an analysis that concludes that CBW production has been significant since the cessation of UN inspections four years ago. Given that regime survival is the ultimate motivation of Iraqi regime policy, it should be anticipated that CBW systems would be used if there were an attempt to destroy the regime.

13.9 A Nuclear Response to Iraqi CBW

To suggest the possibility of nuclear retaliation to Iraqi CBW use may seem extreme but it is relevant here to point to the unexpected outcome of the *Global 95* exercise at the US Naval War College in July 1995, which was designed to explore the development of just such a major crisis. These large 'war-gaming' exercises use personnel from all the armed forces, intelligence agencies and civilian departments to create a detailed assessment of possible conflicts, and *Global 95* was based on two simultaneous crises, in Korea and the Persian Gulf.

In the exercise, the Korean crisis was terminated with difficulty but the Gulf conflict, involving a resurgent Iraq, escalated to the point where Iraq used biological weapons to devastating effect against military forces and civilians in the region. The United States responded with a nuclear attack on Baghdad, ending the war.

The development and outcome of *Global 95* was reported in some detail in a Washington-based defence journal, commenting that:

[17] In 1996, the US Department of Defense made available on the Internet a large number of reports and studies relating to the Gulf War. These included, by mistake, a classified report relating to the National Intelligence Estimate of November 1990 that had concluded that moves to destroy the regime would be likely to entail a biological weapons response. This report was quickly removed from the website but not before it had entered the public domain. Theresa Hitchens, 'Wargame Finds US Short in Biowar', *Defense News*, 28 August 1995.

> The United States has virtually no response to the use of such potentially devastating weapons other than threatening to use nuclear weapons, a Joint Staff official said Aug. 22. But it is unclear whether even nuclear weapons would provide a deterrent, unless the US was willing to take the difficult moral step of destroying a city, he said. On the other hand, if the United States did launch a nuclear attack in response, "no country would use those weapons for the next 100 years," the official said.[18]

Such a view is wrong – a more likely outcome would be paramilitary attacks, in later years, on US cities that would be far more devastating than 11 September, but the point of the Global 95 experience is that the escalation was unexpected and, within the rules of the exercise, apparently not subject to control.

There is a natural reluctance to even consider that the United States, on its own or in coalition, would consider using nuclear weapons, yet it is known that such threats were made at the time of the 1991 war. It now appears that the only circumstances in which the Iraqi regime would have considered using chemical and biological weapons was if the regime itself was faced with destruction. This is precisely the intended aim of the Bush administration, so an Iraqi use of CBW should be expected, including the possible use of biological weapons against civilian populations.

It is also relevant that, since the 1991 war, the United States has developed and deployed a tactical nuclear earth-penetrating weapon intended precisely to destroy deep underground targets such as command bunkers and chemical and biological weapons stores.[19] This weapon, the B61-11, can be deployed on the B-2 stealth bomber, some of which are to be deployed to Diego Garcia for use in military action against Iraq. When Tam Dalyell MP asked the Defence Minister Adam Ingram whether the B61-11 was to be deployed to Diego Garcia, the minister refused to give an answer.[20]

---

[18]Partly because of difficulties in destroying deeply-buried military facilities, the United States has developed an earth-penetrating version of the B61 tactical nuclear bomb, the B61-11. This can burrow 15 m below ground before detonating, producing an earthquake effect. The B61-11 is reported to have an explosive yield that can be varied between 0.3 and 300 kilotons (the Hiroshima bomb was about 13 kilotons) and its principal delivery vehicle is the B-2 bomber. It is therefore highly flexible and fits in with the idea that small but very accurate nuclear weapons might be used in limited conflicts. See "Nuclear Bomb Passes Final Drop Test, Jane's Defence Weekly, 1 April 1996, and Greg Mello, "New Bomb, No Mission" *The Bulletin of the Atomic Scientists*, May/June, 1997. The development of a more sophisticated earth penetrator, the Robust Nuclear Earth Penetrator (RNEP) is awaiting authorisation under the 2003 national defense authorisation bill, see: Eric Miller: "Bunker-Busting Nuke Expands U.S. Options", *Defense News*, 16 September 2002.

[19]Responding to Mr. Dalyell's question, Mr. Ingram said: As he knows, it has been the practice of successive governments never to confirm or deny the presence of nuclear weapons at any particular location or at any particular time. I have no intention of deviating from that line… *Hansard*, Vol. 390, No. 187, column 151, Tuesday 24 September 2002.

[20]The term "weapons of mass destruction" tends now to be applied in an undifferentiated manner to chemical, biological and nuclear weapons. Both of the former can cause heavy casualties, especially if used against unprotected civilian populations. Nevertheless, nuclear weapons can be hugely more destructive and their use would break a threshold that has held since the attack on Nagasaki on 9 August 1945.

Given that it appears likely that substantial UK ground forces will be committed to a war against Iraq in coalition with US forces, it is also relevant to consider UK attitudes to nuclear use. Despite their repeated negative security pledges in conjunction with the Non-Proliferation Treaty, both states maintain policies of nuclear first-use, together with a willingness to use nuclear weapons against states that have other lesser forms of weapons of mass destruction.[21]

13.10 The British Position on Nuclear Use

Since the 1950s, Britain has deployed nuclear weapons on many occasions outside the immediate NATO area of Western and Southern Europe and the North Atlantic. This has included the Middle East, South-East Asia, the Indian Ocean and possibly during the Falklands War.[22] This raises the question as to whether sub-strategic, or indeed strategic, Trident warheads, might be used independently.

Britain reserves this right, and one of the more detailed assessments of the range of options for sub-strategic Trident warheads was made in the authoritative military journal *International Defense Review* in 1994:

> At what might be called the "upper end" of the usage spectrum, they could be used in a conflict involving large-scale forces (including British ground and air forces), such as the 1990-91 Gulf War, to reply to an enemy nuclear strike. Secondly, they could be used in a similar setting, but to reply to enemy use of weapons of mass destruction, such as bacteriological or chemical weapons, for which the British possess no like-for-like retaliatory capability. Thirdly, they could be used in a demonstrative role: i.e. aimed at a non-critical uninhabited area, with the message that if the country concerned continued on its present course of action, nuclear weapons would be aimed at a high- priority target. Finally, there is the punitive role, where a country has committed an act, despite specific warnings that to do so would incur a nuclear strike.[23]

It is worth noting that three of the four circumstances envisaged involve the first use of nuclear weapons by Britain. There has been concern expressed in parliament that the government has not been sufficiently clear about the circumstances under which British nuclear weapons would be used in post-Cold War circumstances. For example, the House of Commons Defence Select Committee noted, in 1998:

> We regret that there has been no restatement of nuclear policy since the speech of the then Secretary of State in 1993; the SDr.[Strategic Defence Review] does not provide a new statement of the government's nuclear deterrent posture in the present strategic situation

---

[21]For a discussion of UK independent targeting see, Paul Rogers, "Sub-Strategic Trident, a Slow-Burning Fuse", *London Defense Studies No. 34*, Centre for Defence Studies, King's College, London, 1996. The evidence for nuclear deployments during the Falklands War is not conclusive but does come from several sources.

[22]David Miller, "Britain Ponders Single Warhead Option", *International Defense Review*, September 1994.

[23]Transcript from the *Jonathan Dimbleby Programme*, 24 March 2001.

> within which the sub-strategic role of Trident could be clarified. We recommend the clarification of both the UK's strategic and sub-strategic policy.[24]

This was, in part, in response to comments made to the Committee by the then Secretary of State for Defence, Mr. (now Lord) Robertson. He had told the committee that the sub-strategic option was "an option available that is other than guaranteed to lead to a full scale nuclear exchange". He envisaged that a nuclear-armed country might wish to "... use a sub-strategic weapon making it clear that it is sub-strategic in order to show that... if the attack continues [the country] would then go to the full strategic strike," and that this would give a chance to "stop the escalation on the lower point of the ladder".

This statement indicated that 'a country', such as Britain, could consider using nuclear weapons without initiating an all-out nuclear war, and that the government therefore appeared to accept the view that a limited nuclear war could be fought and won. It was evidently not the clear statement that the Committee sought, and it did not indicate the circumstances in which such weapons might be used. In particular, it did not appear to relate to whether Britain or British forces had already been attacked with nuclear weapons, or whether nuclear weapons would be used first in response to other circumstances.

At the same time, there has been no evidence to suggest that Britain has moved away from the nuclear posture of the Cold War era that included the possibility of using nuclear weapons first. On 20 March, 2002, the present Secretary of State for Defence, Mr. Hoon, was questioned by members of the Select Committee and appeared to indicate that Britain maintained this policy. In relation to a state such as Iraq he said that "They can be absolutely confident that in the right conditions we would be willing to use our nuclear weapons."

This exchange did not make clear whether this would be in response to a nuclear attack initiated by a state such as Iraq, but Mr. Hoon was questioned on this point on 24 March on the Jonathan Dimbleby Programme on ITV. He was asked whether nuclear use might be in response to non-nuclear weapons such as chemical or biological weapons. He replied:

> Let me make it clear the long- standing British government policy that if our forces, if our people were threatened by weapons of mass destruction we would reserve the right to use appropriate proportionate responses which might...might in extreme circumstances include the use of nuclear weapons.

Later in the exchange, Mr. Hoon made it clear that he could envisage circumstances in which British nuclear weapons were used in response to chemical or biological weapons. He was later asked by Mr. Dimbleby:

But you would only use Britain's weapon of mass destruction after an attack by Saddam Hussein using weapons of mass destruction?[25]

Mr. Hoon replied:

---

[24]Ibid.

[25]Ibid.

> Clearly if there were strong evidence of an imminent attack if we knew that an attack was about to occur and we could use our weapons to protect against it.[26]

The meaning of this is clear enough – there are circumstances where Britain would consider using nuclear weapons in response to a non-nuclear attack involving chemical or biological weapons and would even consider using nuclear weapons to pre-empt such an attack. It may be argued that such a statement is intended primarily as reinforcing deterrence by making the threat more explicit. Although this may be the case, there is the remaining problem that war-time circumstances can provide highly unpredictable environments and unexpected escalations. A culture in which nuclear use is even remotely contemplated is, to say the least, unfortunate.

From what is known about past Iraqi CBW deployments and the likely availability of CBW systems to the regime now, these may well be the circumstances likely to arise in the event of a war.

13.11 After the War

The war itself carries a range of major risks, most of them related to the determination of Washington to destroy the regime and the even greater determination of the regime to survive. Even if the US action succeeds, though, there is no guarantee whatsoever that the consequences will be as expected. Given the widespread existing antipathy towards the United States that exists throughout the region, the idea that a client regime in Iraq would be stable is dubious in the extreme.

In the very best case scenario for Washington, a quick war with few casualties and little economic aftermath terminates the regime and replaces it with one that is acceptable. For reasons given above, this is unlikely but it is possible. But if this were to happen, it would confirm the views of all of those throughout the region who see control of Arab oil as a requirement for Washington. In the short term it might appear successful, but in the longer term it would be an utter gift to al-Qaida and similar organisations, confirming all that they have been saying for more than a decade.

Thus, a war with Iraq should be expected to be hugely costly in human and economic terms and to carry with it the greatest risk of escalation to the use of weapons of mass destruction since the Cuba missile crisis of 40 years ago. Its aftermath, even if apparently 'successful' from Washington's perspective, would entail the development of further opposition across the Middle East to what would be seen as foreign control. It would be directly counterproductive not only for the region but even for the United States itself.

[26]Discussed in another paper in this series: Scilla Elworthy and Paul Rogers, "The 'War on Terrorism': 12-month audit and future strategy options", *Oxford Research Group Briefing Paper*, September 2002.

A counter argument may be made that the intention of the Bush administration would be to aid Iraq in a substantial programme of state-building leading to a democratic government. This is implausible for three reasons. Firstly, the Bush administration does not, in general terms, see it as a function of government to aid state-building in the aftermath of conflict. Secondly, there is recent specific evidence of a failure of the administration to engage in such action in Afghanistan, not least in its refusal to countenance the expansion of the International Security Assistance Force outside of Kabul. Despite the rather more positive efforts of some other countries, the situation in Afghanistan is now highly unstable. Warlordism is rife, opium poppy production has increased substantially and the interim government has experienced the assassination of a cabinet minister and a vice-president and at least two attempts on the life of President Karzai himself.[27]

Finally, the fractured make-up of Iraq, with Kurdish, Sunni, Shi'ite and Christian elements make it unlikely that a stable government would form with ease, particularly one that ensured that the US had full access to Iraqi oil supplies. Such a government would necessarily be seen in the region as, at the very least, a client regime, and more likely a puppet government.

Taken with the current and very widespread perception in the region of Israel as a client state armed largely by the United States, and of Saudi Arabia controlled by an excessively wasteful and wealthy neo-feudal elite, a further increase in the anti-American mood in the region and consequent support for oppositional paramilitaries such as al-Qaida is likely to be the longer term consequences of an enforced regime change and possibly even a military occupation of Iraq.

It is also possible that a paramilitary movement could develop from within Iraq. While there is abundant evidence of the unpopularity of the Saddam Hussein regime, it is certainly possible that internal opposition to US occupation and the subsequent installing of a client regime would result in an evolving insurgency. Internal opposition to the current regime does not equate with the future acceptance of foreign occupation.

13.12 Further Possibilities

Before drawing some conclusions, it is appropriate at least to make mention of some less likely but possible consequences of a war:

- The Saddam Hussein regime may have organised its own paramilitaries, or may assist surrogates, to conduct a range of actions against US and other coalition forces. These could well include, for example the use of chemical, biological or radiological weapons against rear supply bases in the region, or transit routes such as the Suez Canal.

[27] See, for example: Patrick E. Tyler and Richard W. Stevenson, "Iraq war could shock shaky U.S. economy", *International Herald Tribune*, 31 July 2002.

- The regime may even choose to take an action from which it has so far refrained – providing a surrogate group with chemical or biological agents for use against targets in the United States, or against US facilities in Europe, or against states supporting the US in its war on the regime. In this regard, the decision of the British government to vaccinate health professionals against smallpox is an indication of its perception of the risks that might arise from the war.
- Attacks against oil tankers and other aspects of the oil and gas supply chain may be mounted, possibly using surrogate paramilitaries, with the hope of affecting the price of oil.
- The regime may choose to engage in a small-scale action against Israel in order deliberately to incite the Sharon Government to enter the war, confident that this would greatly increase regional support for Iraq.
- The Israeli government may use the cover of the war to engage in much harsher action against Palestinians in the occupied territories, including forcing many of them into Jordan.
- In such circumstances, and given that Hezbollah militia in Southern Lebanon have recently received some thousands of short-range missiles from Iran via Damascus, Israel might suddenly find its northern cities under attack and would respond with forceful counteraction against militias and Syrian forces in Lebanon.
- In the extreme circumstances of imminent regime destruction, the regime may choose to act against Israel in order deliberately to bring about Israeli retaliation with the regional reaction that this would bring.

More generally, there is a substantial perception in western business circles that a war with Iraq could precipitate a recession.[28] This is a matter of particular concern in the civil aviation industry where the effects of the September 2001 attacks have still not been overcome.[29]

13.13 Consequences of a War

This briefing has explored the probable features of a war with Iraq, and it leads to three substantive issues that need to be taken into account in relation to such a war – civilian casualties, the risk of escalation to the use of weapons of mass destruction, and the question of regional stability following enforced regime destruction.

---

[28]"CEOs Leery of Iraq War", *Aviation Week and Space Technology*, 30 September 2002.

[29]Philip H. Gordon, Martin Indyk and Michael E. O'Hanlon, "Getting Serious About Iraq", *Survival*, Vol. 44, No. 3, Autumn 2002, pp. 9–22. According to this analysis: "If elite Iraqi armed forces fight, then, Iraq could suffer 10,000 or more military deaths and a similar number of civilian deaths, while US troops might lose from many hundred to several thousand soldiers in action" (p. 18).

From what is known of the likely process of the war, one should expect that there will be a very strong probability of urban warfare, especially in greater Baghdad, and that this will involve determined efforts by the regime's elite forces to ensure that US and other troops fight in a confined environment in the midst of a civilian population.

There is credible evidence that over 3,000 Afghan civilians died directly in the war last year, besides many more thousands dying in refugee camps. The Israeli experience in Beirut in 1982 also indicates the level of likely casualties. A recent assessment by analysts at the Brookings Institute in Washington, published in *Survival*, the journal of the London-based International Institute for Strategic Studies[30] considers that Iraqi military casualties might be in the order of 10,000, with a similar number of civilian casualties. The authors do not assess this in relation to the possible use of chemical and biological weapons and do not consider the possibility of further escalation. Their assessment is in relation to a conventional conflict.

Such an assessment may well be conservative, not least in relation to some of the issues discussed here, but it is entirely reasonable to conclude that civilian casualties in a war with Iraq will be at least 10,000, more than three times greater than the number killed in the attacks of 11 September 2001.

Furthermore, this paper does not seek to examine the longer term effects of a war on the Iraqi population, although evidence gained after the 1991 war suggests that post-war civilian deaths included 30,000 refugees, at least 4,000 non-refugee deaths and a doubling of child mortality rates.

On the second issue of the use of weapons of mass destruction, this report leans to the conclusion that Iraqi use of chemical and biological weapons is highly likely given the fundamental requirement for regime survival. How such weapons might be used against a sophisticated and protected enemy and with what effect, is more problematic.

At a minimum, their use should be expected to relate to a war aim of forcing coalition troops to be encumbered by the need to fight in protective gear, but the regime may, for example, employ most of its weapons strategically, including attacks on rear bases, logistic supplies and even centres of population, with the consequent risk of heavy civilian casualties.

If such CBW attacks were successful to the point at which they caused substantial casualties, then scenarios such as that explored in *Global 95* are relevant, including a coalition nuclear response. Any rational analysis would conclude that this should not be the outcome, and it may not be, but UK government statements, in particular, are remarkably frank in this connection. If the government is ready to

[30]*Counting the Human Cost of the Gulf War*, Medical Education Trust Background Paper, London July 1991. At the time of writing this briefing, an assessment is being undertaken by the London-based non-government organisation MEDACT into the longer-term health consequences of a new war on a population already affected by a decade-long sanctions regime. This study, *Collateral Damage: the health and environmental costs of war on Iraq*, will be published in mid-November 2002 and will be available at www.medact.org.

embark on a war in which it is prepared to consider the use of nuclear weapons this should, at the very least, be subject to intense public discussion prior to a war.

Finally, there is the question of the aftermath, and here the analysis suggests that the most likely effect of a war will be to increase opposition to what will be seen as increased US control of the region. This is likely even in the best-case scenario of regime termination with minimal violence. The more likely consequences of a war – substantial immediate civilian casualties and longer term socio-economic damage – are likely to increase such antagonism still further. This, in turn, should be expected to lead to an enhanced environment for paramilitary organisations to promote their policies and actions, both in the region and against the United States and its allies.

All of these issues are indicative of the dangers inherent in the decision to terminate the Saddam Hussein regime by the use of force. At the very least they demonstrate that it makes eminent sense to investigate and develop policies that present a viable alternative to war.

Chapter 14
Letters from Raqqa (2014–2016)

14.1 Introduction

Ten "letters from Raqqa" have been published in *openDemocracy* covering a period of a little over two years, the most recent one being on 8 December 2016. They are part of my regular series of weekly articles which have been published since September 2001, offering close analysis of the "war on terror", but with a difference: they are written as if coming from an ISIS supporter working in Raqqa, northern Syria.

The letters therefore give a view of the conflict which contrasts with the typical analysis from western perspectives.

In doing so they use a wide variety of sources, both published and oral, to try and get inside the thinking of a well-educated, intelligent but utterly committed ISIS supporter.

Reaction has varied greatly. Some deeply antagonistic and vituperative comments angrily condemn the letters as unalloyed ISIS propaganda. Many others have been much more positive, welcoming them as a necessary attempt to examine perspectives that are radically different to the norm.

This level of interest, in the context of a conflict that shows every sign of evolving further in 2017, suggests that it might be useful to gather the letters together. They appear below, each with a brief introduction to set the scene.

14.2 First Letter (9 October 2014)

The first letter introduces the 'writer' and his family, the impact of the war, and his motives for joining the group and moving to Raqqa. It refers to the start of the air war three months earlier, while celebrating the successes of the movement across northern Iraq earlier in the year as a model for long-term success

P. Rogers, *Paul Rogers: A Pioneer in Critical Security Analysis and Public Engagement*, Pioneers in Arts, Humanities, Science, Engineering, Practice 21,
https://doi.org/10.1007/978-3-319-95150-8_14

When I came to Raqqa two years ago my brother had been here for more than a year and was already a platoon leader. He had fought the American special forces in Anbar for three years until captured and tortured late in 2006, so he had even survived Operation Arcadia. After four years in Bucca camp – a singularly formative educational experience where he learnt much about US military attitudes – he was finally released. He had, though, lost two uncles and three cousins to American and British attacks, with our father, two aunts and four cousins killed in airstrikes and two more cousins maimed for life. It was hardly surprising that he joined up with the Baghdadi group at the first opportunity, nor that I should follow him not long afterwards.

So where are we now? First, let's just remember that we have more than one aim and a thorough mix of motives. Behind it all is the desire to be part of the historic mission to restore the Caliphate, bringing true rule to the world even if it takes centuries, but there is more to it than that.

Inevitably we have hugely bitter feelings towards the Shi'a in Iraq and do not see Haider al-Abadi as any different to Nouri al-Maliki. We are passionately opposed to Bashar al-Assad and his Alawi clan but also hate the utterly unacceptable Sunni regimes dominating our region, most of all the appalling House of Saud and its preposterous claim to be Guardian of the Two Holy Places. Beyond that lies the American-Israeli nexus, its occupation of the Third Holy Place, with the Zionist massacres of our Palestinian friends a continuing atrocity.

I have only been here a couple of years and much of my work is analysing western attitudes, but I simply cannot understand the reaction to our killing of hostages, Iraqi soldiers and police and other opponents. Is it really worse to be beheaded, a near instant death, or to die slowly from hideous burns, from being shredded by hundreds of sub-munitions or crushed slowly to death under a collapsed building? Sure we make an example of captured soldiers and police, but that scares off others and makes our task easier. It is no different to innumerable attacks done by the Americans over the past ten years, including those which they themselves characterised as punitive.

Yes, we dress hostages in orange jump-suits and maybe that makes a few in the west think, but how many know there are still more than a hundred in Guantanamo, many there for more than a decade and likely to be kept in prison for the rest of their lives – a slow living death with no hope of an end to it, not quick and merciful.

But I'm getting away from myself when I really wanted to write down where we are now, perhaps the start of an occasional diary but I will keep this first entry brief.

The Americans started their airstrikes exactly two months ago, more or less when we expected them to do so, and so we had dispersed most of our assets in good time. They did have some effect here and there and halted a couple of advances, but much of the equipment they hit was surplus to our requirements and was mostly their own stuff anyway.

They've done 388 raids so far, the majority in Iraq, and they are now having trouble trying to identify suitable targets. Yes, they do have British, French, Belgian, Danish, Australian and Canadian aircraft either here or on their way, as well as some from a few local states, although the latter are almost entirely incompetent.

They still say they will not put "boots on the ground" but the numbers of special forces are creeping up and the Pentagon is currently moving an army brigade HQ to Iraq, a sure sign that more is planned.

The only operation that took us by surprise was the first big raid on Syria, but it was not their attacks on us that caught us out. Apart from a few unlucky supporters killed at a roadblock outside the city we had long since evacuated everyone and had also dispersed almost all our weapons and kit.

No, the surprise was the effort they put into trying to damage the Khorasan group over in the west. We have nothing to do with them and their weird insistence in plotting attacks abroad, but they certainly scared the Americans. In the event, it kept the pressure off us so Khorasan actually did us a service.

I also find it weird how the western press seems to pick on just one happening – the Mosul dam attack or, just now, the fighting around Kobane – not getting even a remote understanding of the wider picture. What that picture shows is that we are making progress on many fronts, especially but not only in Anbar province. Even in Syria, Assad's crowd still see us as an asset and long may that last, while the Turks are so conflicted on the Kurdish question that we see little threat from them apart from a possible symbolic action.

Our progress across northern Iraq three months ago was no surprise to us – after all we had been preparing for it for more than a year, and we are not seriously affected by the air attacks. We are therefore in a good position to consolidate our territorial control as we prepare for the long war. Indeed, our morale after two months of attacks has never been higher, born from the pleasure that we are now engaged in combat with a serious enemy not joke forces like the Iraqi army.

We are also doing amazingly well with all our international communications. The sophistication of the media operations is a joy to behold and is way ahead of anything that the enemy can muster, especially in reaching out to young believers and drawing them to the cause. I understand that the leadership has some worries about the force with which mosque leaders are condemning us in some countries such as Britain, but these are small setbacks in an otherwise positive environment.

There is much else I could discuss but perhaps I might end on one big issue – where next and what do we most want our enemies to do? I am not close enough to the leadership to be sure, but the city is a hotbed of gossip and what I try to do is to listen to those sources that have been accurate in the past.

The word there is that what is most wanted is serious numbers of western boots on the ground. There are scores, if not hundreds, of men who fought Task Force 145 – a Rangers battalion, SEAL Team 6 and an SAS squadron – especially in that crucial 2005–06 period, and they seriously want revenge for the wholesale slaughter and torture of their friends and relatives.

Indeed, many of them live in eager anticipation of the opportunity to capture western troops and then dress them up in the orange suits, waterboard them, and execute them, all on video for worldwide distribution. Revenge will be sweet and for quite a few of them revenge is a far stronger motive than seeking the new

Caliphate. It is just as well that the Americans and British have no understanding of this, making it all the more likely that they will blunder into yet another trap.

How will this trap be sprung? Difficult to say, but I know one of the senior people close to Baghdadi is obsessed with the Tet offensive nearly half a century ago, sure that this time it will make the Americans increase their forces rather than withdraw. Where will it be sprung? Again, I can't be sure but there is one pointer. The western media is making quite a lot of our recent advance between Fallujah and Baghdad, including this week's success in and around Abu Ghraib.

What they all seem to be missing is that we already have highly effective yet dispersed forces well ensconced in the western districts of Baghdad itself. Take a look at the area around Baghdad international airport, not doing much on the civil side but an absolute hive of military activity as the American forces pour in (Bing is far better than Google for this – much clearer maps). Then look to the west and east of the airport complex. Westwards is Abu Ghraib, barely ten miles away with lots of farms, irrigation ditches and villages between it and the airport – quite decent paramilitary country and easy to disperse. Then look to the east and within five miles are the crowded Baghdad suburbs of Saidya, Khadra, Ameria and, of course, the appropriately-named Jihad. None of them is openly under our control but we are everywhere.

I may be wrong, but this is one bit of gossip I take seriously. Watch this space as we do our very best to wreak havoc and then get the Americans really involved, and this time on our terms.

That's all for now, more in a couple of months.

14.3 Second Letter (31 October 2014)

A second letter, written soon after the first and giving more background. Somewhat controversially it implies that ISIS has quite a sophisticated intelligence and analysis capability, which did not go down well with some readers of the column when it appeared in openDemocracy. However, it is a reasonable assessment given the more widely accepted view that ISIS ran, and still runs, a very professional if often appalling news dissemination programme via the web and new social media.

When I last wrote I said that I might add something in a couple of months, but the questions you raise have prompted me to make a more immediate response. The first two were: what am I doing here just now, and how did I come to be doing it?

As you know, I came two years ago to join my brother and fight for the cause of an Islamic Caliphate. My motivation, as was his, was primarily revenge, given that we had lost two uncles and three cousins in fighting the Americans and British, and our father, two aunts and four cousins to airstrikes. That may still be part of our motivation, particularly the death of our beloved father, but we now see a much more positive future as we embrace the prospect that our leaders hold before us of a true Islamist entity. Whether we live to see it in this life is not relevant – that we are already part of it is.

My original journey here, my haphazard training (quite unlike the professionalism we have now) and my induction into fighting were all over within four weeks when I was caught in a Zionist attack, losing my left arm and very nearly my life. I survived, recovered and was desperate to return to the fight, but our leaders had other plans, telling me bluntly that I could play a far more important role for our cause by joining the analysis team SOBRA (I understand that this stands for State Office Briefing Room A, the place in our main bunker where we originally worked).

After more than eighteen months into this work, I have to accept that they were right – it does make far better use of my Masters degree from SOAS and my three years of living in the UK and USA, and I now lead a small team that monitors western media and government output to prepare briefings for the leadership. I have three people working for me, and our whole section numbers more than twenty, covering all the major western languages as well as Chinese and Russian, and with excellent communications systems that have so far been entirely unaffected by the numerous US airstrikes.

Most of our output is for the main planning cells, with some of it going right through to the leadership. But we also feed in a constant supply of information to our colleagues in media production. They tend to use our material in a highly nuanced if not frankly propagandistic manner, but I have to admit that when it comes to propaganda they are the very best, and simply streets ahead of their western opponents. Their numbers have increased substantially and there are now over thirty of them, many being recent recruits from among the more knowledgeable of our western brothers and sisters.

That, incidentally, is an area where the rate of expansion is hugely positive. We now have many thousands of young recruits joining us from across the region. Even more importantly, many hundreds a month come from western countries, mostly men but with an increasing number of women.

You ask how I think the struggle is going, especially with what outsiders see as our failure to take Kobane. I have to say that our leaders have little concern, for two reasons. First, we fully expected that at some stage the Americans would try to start a serious air war and would eventually strong-arm the weak Turks to allow the Iraqi Kurds to help defend the town. Both are proving to be useful training exercises for our less experienced militias.

Second, as you will recall, our core military leadership has many people who learned how to handle the Americans in Iraq eight to ten years ago, but we have thousands of younger fighters with far less experience. This is what they are now getting. It is going to prove invaluable during the coming winter when the Americans will really step up the air attacks against us here in Raqqa.

One of the things we are expecting is a determined and sustained effort to wreck our civil infrastructure. Transport and communications will be the priority, together with the sustained disruption of power supplies. One of my recent assignments was to investigate the current status of the American "blackout bomb" that they used in Serbia in 1999, disrupting power supplies of over 70% of the country. I've found out that it is very much around, designated the BLU-114/B, and we expect it to be used frequently this winter, so much so that our leaders are already preparing counter measures. You have probably never heard of this, so here's a link.

You also ask me about morale and I can only reply that it is currently very high. As I have said, my main function is to analyse the western media and I must admit that they still have little conception of what they are dealing with. They report, almost jubilantly, our failure to take Kobane but cannot understand that this is little more than a sideshow. Meanwhile, they miss out so many other developments.

Our mission is to create a new Caliphate, starting here in Syria and Iraq but spreading out over the next decade or more to bring in links right across the Islamic world. Let me just give you just three examples of current progress.

First, our leaders have now formally stated the connection between our cause and the suffering of our Uyghur cousins in China. Just making that statement, and publicising it widely across our world, begins the process of unification.

Second, as the Americans and British finally withdraw from Afghanistan, our Taliban cousins spread their control over more and more territory. They are doing it quietly but to great effect and this will continue, with a substantial increase in control after the winter. We do not pretend that we control them, nor do we seek or need to do so. In the wider scheme of things it is enough that they make progress.

Third, I simply cannot understand how the west, especially the Americans, fails repeatedly to recognise the effect of the actions of the Zionists. Even now, they have no appreciation of how useful the Gaza war was to us and how much anger it induced across the Muslim communities in the west – and still does as the Zionists and the Egyptian leader al-Sisi together block the rebuilding.

On top of this, Binyamin Netanyahu announces the building of 1,000 new settler homes in East Jerusalem and then closes Haram al Sharif, one of the holiest of all our sites. It is simply unbelievable, and the effect on recruitment to our cause will be a joy to behold. Do you seriously need to ask how we see the future? To say that we see it with confidence is a masterpiece of understatement.

14.4 Third Letter (22 January 2015)

Nearly six months into the air war and his attitude is rather less exuberant, admitting to considerable losses and even suggesting that the leadership might have underestimated the impact. Even so, he remains very committed and outlines why he thinks the ISIS paramilitaries are so resilient in the face of the intense air attacks. He also explains the function of the Charlie Hebdo attack in Paris and the manner in which ISIS aims to stir up anti-Islamic antagonism as a substantial part of its strategy.

I had hoped to write to you before the end of the year but, as you may have heard, my brother was badly injured in an American airstrike and I spent a lot of time making sure he got the right treatment. The attack was on the Fallujah-Abu Ghraib road and while he got immediate medical help there, he had to be brought back to our much better facilities here in Raqqa. I am really pleased to say that he is recovering well and very anxious to get back to his platoon.

You mentioned in your last message that you had passed on my letter to some friends, one of whom scanned it and put it online and, as a result, it even got picked up by one of those western websites that quaintly thinks it presents a broad spectrum of opinion on our region! Never mind, perhaps someone will read it and get the message, but since you might want to do this again, I'll just report a bit of background.

My brother came to join the cause three years ago after many members of our family had been killed, and I joined him a year later. I had come to fight but got injured early on, losing an arm. They then had me join the group analysing western attitudes and their media, seeing us through their eyes to help our planner conduct the war most effectively. I am part of a large group working in many languages but because of my SOAS degree and years in London, I concentrate on Britain.

When I last wrote at the end of October, the air war was already under way and we had experienced nearly 400 attacks, losing some of our soldiers very early on. Fortunately our losses were not too great and, in any case, are mostly new recruits – martyrs are serving as a marvellous inspiration to young people to come and join us. Now that the war with the far enemy has really started it is so much easier for us to present the case that the Islamic world is once again under attack from the Crusader-Zionist conspiracy, and we can point to all the other examples – Afghanistan, Iraq, Yemen, Pakistan, Somalia, Mali, Libya and the rest.

Already we are seeing an increase in the numbers of young men and even women coming to our cause. I saw a recent western estimate of 18,000 joining us so far, including 3,000 from outside the region. Surprisingly, they are not too far out with these figures except that they are seriously underestimating the numbers from western Europe.

You ask how we are coping with the attacks. Well I have to say that the intensity has caught us just a little by surprise. There have been about 3,500 attacks so far on nearly 2,500 targets, with around 7,000 bombs and missiles dropped. Our statistical bureau colleagues tell us that over a thousand of our fighters have been killed and about the same number injured, but that our paramilitary leaders were planning for much higher losses and are not remotely worried – we and they may mourn the deaths but celebrate the martyrdom.

There have been many civilians killed, including children, and our social media people have done a remarkable job in telling their stories and sending the news right around the world. This, alone, is a really helpful recruiting aid.

Where does it all leave us, and you rightly ask how the war is going? As you know I am not in the propaganda business and in the analyses I organise for the leaders they insist on me telling them it as it is, so I will be straight with you, too. Firstly, our progress has been blunted by the attacks but no more than we expected. We are not even bothering to put resources into taking Kobane, as that had little strategic importance being more a matter of presentation than anything else. Secondly, we have actually made some progress in Iraq, especially consolidating our hold over large parts of Anbar province and increasing our infiltration of western Baghdad. Finally, and most importantly, we have made significant territorial gains in northern Syria, aided by the agreement with al-Nusra which continues to hold in most districts.

You may wonder how this can be when we are facing an air assault that is much more intensive than any meted out to the Afghans in recent years. The main reason is that our paramilitary leaders know what to expect since most of them fought the Americans and their allies for years in Iraq. One of my friends in military planning, who fought in Chechnya, Kashmir and Afghanistan before being injured in Iraq and transferring to desk duties told me that in his considered opinion our paramilitary leadership is probably the best in the world, and certainly superior to the other side. With all their advanced weaponry and firepower, they are not remotely near understanding us or what we are about.

As I mentioned, my duties involve analysing the western media and I am required to pay particular attention to the military literature, especially in the United States. What I do see there, if only occasionally, is some public acknowledgment that they are facing a much more difficult task than they imagine, and that unless they can train effective Iraqi army forces this war will last for years. Even so, there is very little attention paid to such views.

Interestingly, I occasionally see reports from well-informed analysts in civil think-tanks pointing out that generational cohorts of combat-trained paramilitaries have been produced, first in Afghanistan in the 1980s and again in the 1990s, then in Afghanistan, Iraq and Yemen, more recently in Libya and now once again in Iraq and Syria. It is hardly surprising, is it, that the current generation is an elite in every sense of the word? Fortunately, the very few analysts saying this are systematically ignored, which is just as well for us.

So what of the next few months? Well, three elements are significant. As I said last time, we are very keen to have enemy boots on the ground fighting us directly. We are now getting this – the recent engagement with Canadian special forces is an example. The more of this the better, and what we really want is ordinary troops involved. We can then capture some, build up a pool of prisoners and execute them one by one, no doubt ensuring a substantial escalation in the war.

That is the second point: that outcome is precisely what we want. I still cannot understand that the leaders of the far enemy do not appreciate that we want a war! Do I really have to shout it – for it seems so obvious. If our main near term aim is to show we are defending Islam, how can we do that if we are not being attacked? It would make no sense. How can we ensure an increased supply of recruits if there is little work to recruit for? As one of their TV adverts puts it – 'simples'!

That brings me to my final point, the impact of the Paris attack and what it means in the coming months. The brothers who staged that attack were not part of us, but they did a very good job of stirring up anti-Islamic antagonism. This is something we really do need if we are to get more disillusioned young people flocking to the cause. In fact if we can ensure that inter-communal relations across the west are wrecked then that is really good news.

Because of this we have developed over the past year a new department that works specifically to place dedicated young people with direct combat experience back in their own countries in order to prepare for attacks. We are obviously losing some to the security forces but plenty are surviving. We anticipate a number of incidents in the coming months and have some brilliant targets identified which will come as great surprises and have maximum effect. As well as directly inspiring more recruits to our cause, far more of our Muslim friends will be more marginalised and alienated by the fracturing of society that our attacks cause, with ever more of them coming to see that ours is the only way.

In short, we can survive the air war with ease, for months if not years. We welcome and hope earnestly for ground combat. And we look forward to more strikes at the heart of the far enemy and the sense of mission that will be engendered. In short, things are going very largely according to plan.

I will write again with news of my brother's progress in a few weeks but I very much suspect he will be back in action long before then – after what he has suffered, and all the losses to his family, there is just no stopping him.

14.5 Fourth Letter (22 March 2015)

Two months later he presents a still-positive account of where ISIS is, what its strengths are, and his insistence that it is an entirely positive cause. As is usually the case, he sees developments beyond Iraq as frequently favouring the movement, most notably Binyamin Netanyahu's electoral success. To him, and ISIS, the Israeli prime minister is an asset because of his treatment of the Palestinians. His own work reflects this reverse view – he is to run an analysis unit on Britain's forthcoming election, the hope being that right-wing parties will succeed. For ISIS, anything that adds to the alienation of Muslims in the UK is to be welcomed, as is a British government of a hawkish persuasion.

Thank you for asking after my brother. When I last wrote I was confident that he would soon be back on the frontline, probably in Tikrit, but I am sorry to say that one of his wounds turned septic and for a week we thought he would lose his right leg. Fortunately his brigade commander heard what was happening and pulled sufficient weight for him to be brought back to one of our military medical centres here in Raqqa. The treatment has been excellent and he is now up and about. One of the good things about this is that I have been able to see him almost every day and we have been able to talk and share our visions of the future more than at any time since I first came here over two years ago.

You ask about how I see the war going and I will tell you, but first let me bring you up to date on my own news. You will recall that I originally came here to fight, following my brother and utterly determined to aid the cause after the huge suffering our family had experienced – my brother and I still talk fondly about our beloved father and his terrible death in that crusader airstrike.

My fighting life ended early when I lost my left arm in a Zionist air attack and since then I have worked for the leadership in the SOBRA team, my responsibilities being to monitor and analyse the western media, especially in Britain. They seemed to recognise my hard work and my ability to be ruthlessly independent in my analysis and I was promoted to run the whole unit early last year. You might expect that they don't want to be presented with bad news but that is simply not the case. They are so hardened, not least from the experience of many of them against those crusader special forces in Task Force 145 in Iraq, and they are absolutely insistent on being told it as it is.

One of their increasingly significant requirements is for high-quality analysis of the political trends in those crusader countries that provide us with such dedicated recruits. Britain and France are important although they were also particularly concerned about Israel when it looked as though Binyamin Netanyahu would lose the election – a potential disaster for us since he has been such a marvellous recruiting sergeant for our cause.

The relief at the result right across the movement has been palpable. Not only will he encroach still further on our lands but there is every chance of his stirring up a crisis between the crusaders and the hated apostates in Tehran.

It may not happen soon, and there might even be an initial nuclear deal, but we are not in this for the short term so that is hardly important.

But what I really wanted to say is that the leadership instructed me last month to establish an election research unit within SOBRA to provide analysis of electoral trends in key crusader states, with an emphasis on impending elections. My first task, already underway, is to analysis the forthcoming British election and how we might influence it in the right way. I'll say a bit more about that before I finish but first let me respond to your query about how the war is going.

It is two months since I last wrote to you and the short answer is that it is going more or less according to plan, especially in Tikrit. I know that would have surprised you a couple of weeks ago, given the way the Iraqi propagandists had been trumpeting their early progress but I'm sure you will have heard what has happened recently. In effect, the operation has stalled even though we have less than 1,000 fighters, facing 3,000 Iraqi soldiers and 20,000 Iraqi militia fighters supported by many Iranians including their special forces.

The Iraqis are losing at least sixty of their people every day, with scores more injured who are now clogging up their hospitals. For the time being they have stopped trying to advance and are waiting for reinforcements. At some stage we will suddenly withdraw all our forces and they will claim a great victory. We, on the other hand, will have ended up with hundreds more recruits having got first hand combat experience while also taking territory elsewhere.

I'm not sure whether you know that only last week we overran the headquarters of the Iraqi army's 26th brigade at Tharthar, close to Baghdad. We killed or captured many Iraqi soldiers and took truckloads of equipment and munitions before moving on.

There is also a huge advantage coming to us from the actions of the hated Iraqi Shi'a militias as they plunder and burn Sunni villages. This has been a persistent element of the war that has been going on for many months. One example was when a large Shi'a militia force tried to take control of Amerli (or 'liberate' it, as they say) last summer. The fighting was hard and in the process they staged scores of reprisal raids against Sunni villages in the surrounding areas. At least thirty were attacked, the villages looted, men abducted and thousands of buildings burned to the ground.

This has now been happening around Tikrit and we have no doubt that the effect will be to increase support for our cause right across the province, as happened around Amerli. Tacitus had it right when he said "they made a desert and called it peace", and this will be repeated time and time again. In the short term they will make gains, including, no doubt, Tikrit, but in the longer term these will greatly play to our advantage.

We are being helped in many other ways too. Little by little other fighters are rallying to our cause, from Libya right through to Kazakhstan. Also, the museum attack in Tunis last week is having precisely the intended effect. Picking the afternoon of a cruise-ship visit and crusader tourists present there means that it had a worldwide impact, and hitting Tunisia's tourist trade is hugely important. With its rampant graduate unemployment and hundreds of thousands of well-educated young people excluded from mainstream society, Tunisia has long been one of our best recruiting grounds, but the resurgence in the tourist industry and the employment provided was a worry for us. This will now be happily averted, and not only will the government crack down heavily on dissent but there will be more and more angry and frustrated people on the margins.

The attacks on the apostate mosque in Yemen are also very helpful especially in boosting our financial support base in Saudi Arabia. The Saudis have been hugely worried at the rapid increase in Iranian political influence in Iraq, a key part of their much-feared *Shi'a* crescent from the Mediterranean to the Arabian Sea, and their near-paranoia has been greatly boosted by the rise in Houthi power in Yemen with its obvious Iranian backing. For our people to attack apostate mosques to such an effect is a clear indication to our Saudi backers that we are increasingly a force to be reckoned with.

I mentioned earlier on that I would say more about our new Election Research Unit. As must be obvious to any intelligent person, if we are wanting to increase the number of recruits coming to our cause from outside the region, then the more angry, resentful and marginalised young Muslims there are, the better. In the long term we have high hopes for Marine Le Pen and the Front National, even if her claims to being the "first party of France" are currently a bit over the top.

Our immediate concern, though, is with Britain and the forthcoming general election where we have some serious worries, almost as bad as the fear that Netanyahu would lose in Israel last week. Obviously what we want in Britain is the best performance possible from UKIP and Nigel Farage. They are great assets as they play to the fear of immigrants in general and Muslims in particular.

The ideal outcome, which looked plausible a few weeks ago, would have been a weak Conservative Party trying to form a government but dependent on UKIP to form a new coalition. In our dreams the real delight would have been a weak Cameron as prime minister with Farage as deputy prime minister and home secretary (what the British call the interior minister). Just imagine having Farage in overall charge of community relations and immigration – it would have made our day!

Maybe it still will, but the problem is that there is a risk that Cameron will not even be in that position, especially as UKIP looks ominously to be past its peak. There is now a real possibility that Miliband could end up with the largest number of seats and our even greater worry is that he forms a minority government with tacit support from that Scottish lot, the extraordinary thing being that they actually want immigrants north of the border! Can you believe it? We could actually see a government in power that is favourable to minorities, including even Muslims. This really is our worst nightmare and hugely damaging to our growing support base there.

We have already passed on this analysis to the SOBRA leadership and they are putting together plans to try and prevent this, but they have already run into difficulties. One proposal was to make a determined effort to capture some British military who could then be treated as war criminals in the usual way, perhaps in an appropriate manner over Easter.

The problem is that the British military have, for once and rather unusually, seen this coming and have recently minimised their exposure, including postponing new military training missions in Iraq until after the election. This is really annoying as it would have been by far the best way to really stir up Islamophobia in Britain and with it more support for UKIP.

I have no doubt, though, that SOBRA is looking at other ways of affecting the election result and you are likely to see something significant happen in the coming weeks. Even if it doesn't, Britain does look like entering a period of political uncertainty and that, at least, is good news for us – uncertainty means scapegoats and we all know what that involves.

14.6 Fifth Letter (14 May 2015)

He "explains" the loss of Tikrit and points to other successes, as well as highlighting his own work. His view of the British political scene immediately after the general election is positive: the Conservative victory is welcome, as is the continued significance of Nigel Farage, while there's relief that Ed Miliband's Labour Party didn't win (out of fear that Britain might adopt a more cautious stance on military action against ISIS). From his perspective, ISIS wants to be attacked in order to maintain its status as the defender of "true" Islam.

Thank you again for asking after my brother. I am pleased to say that he is now almost fully recovered and expects to return to the fight within a couple of months. For now, though, the leadership has given him a training role. That's something he is really well suited to given his years of combat, especially during the war with Task Force 145 back in 2005–06. His experience under torture at Camp Bucca has also proved a great asset. It gives him a real authority with recruits, and helps him prepare them for what they might face as our war escalates here and in Iraq.

My brother tells me that there is a good chance that he will be sent to northern Afghanistan to help with the further training of Taliban and foreign fighters.

Apparently our recent support for the Taliban in the fighting around Kunduz has proved to be welcomed by their leadership, helping to overcome any suspicion of our motives. At the very least we will be able to work with them some more and our long-term aim is that they embrace our vision of the new Caliphate, a much more potent idea than their own rather limited ethno-nationalistic outlook.

You ask me to let you know what I think of recent western political developments, now that I am full-time on the SOBRA department that analyses external developments for the leadership. I am happy to do so but will first say a little about the recent western insistence that the war is going well for them.

For us in SOBRA we really find it quite laughable. Those people in Washington seem to think that because we made such rapid progress from Fallujah to Mosul a year ago, any slower rate of progress indicates impending failure, Tikrit being a good example. They seem quite unable to understand what we are about, or the long timespan of our programme. Even in the short term, though, they get it wrong, though we suspect that the real reason for protestations of success is to keep their own domestic opinion quiet.

Look at it from our perspective. In the past nine months we have experienced close to 10,000 airstrikes with over 6,000 targets hit, yet are still as active as ever. We only had a few hundred men defending Tikrit and most of them withdrew successfully when the planned decision was taken. Tikrit, though, is now a wrecked town, as was Kobane last year, and one of the key results is that on the few occasions that they take territory, the level of destruction that they leave behind just makes for more converts to our cause as our Sunni brethren return to their shattered homes, shops, factories and farms.

Meanwhile, while they take the occasional town we make progress elsewhere, including that destruction of their brigade HQ the other month. Take a specific example – we currently have less than 200 of our fighters controlling a large part of the Baiji refinery, ensuring that such an important source of revenue is unavailable to their regime. Also, the complex of pipes, tanks and industrial plant make for an excellent guerrilla warfare environment. They cannot bomb it because that would be self-defeating, while their own soldiers are simply not up to taking us on – and the *Shi'a* militias are no better. At some stage they will assemble many thousands of men, as they eventually did around Tikrit, and we will quietly and calmly withdraw and move on elsewhere. We may not even bother to wreck the refinery since we will be back before long.

As far as Baiji is concerned you may ask, why do it? Well the answer is straightforward -symbolism. For us to control one of the country's most important industrial sites for weeks and months with little more than a token force is a vivid demonstration of our capabilities and is not lost on *Sunni* communities across Iraq as they suffer under the apostates. Moreover, it has a similar impact in the wider world and posting that video of our surveying the refinery with our very own reconnaissance drones has had a huge impact. Meanwhile, as I said at the start, we are becoming steadily more active in Afghanistan, a development as welcome as it has been unexpected.

But you ask specifically about my work in election monitoring and I will tell you about that now, especially in relation to the British election result. The unit as a whole puts its greatest emphasis on the United States, followed by the UK and France, but it also looks at others, including Stephen Harper's Canada and dear old Tony Abbott over in Australia. Those last two don't receive much attention at present, although Canada's decision to send its CF-18s into action in Syria as well as Iraq is a thoroughly welcome development, not least as the chance of one of their planes getting shot down and the aircrew captured has risen.

France is of some interest not least because of the continuing and greatly welcome strength of the National Front, and the United States has no elections in the near future. Mind you, we are looking quite positively at next year's presidential election. Any Republican that makes it to the White House is well-nigh certain to be more hawkish than Obama and we have no worries about the hard line nature of a Hillary Clinton administration. Our concern is if a more liberal Democrat makes progress in the primaries, but we will cross that bridge when we come to it.

That leaves us with Britain and I have to confess that a week ago we were getting seriously worried because of the risk that Miliband would somehow get in. Our nightmare was Miliband running a minority government with informal SNP and LibDem support and being both cautious over foreign wars and careful about how to tackle what they call 'extremism'. Both would be singularly unhelpful to our cause. You will recall that in my last letter I said that our dream result would be a Tory coalition with a vibrant UKIP and Nigel Farage as deputy prime minister and home secretary – manna from heaven (if you will allow me to borrow a Crusader phrase!).

In the event we actually think we have done better than this since a Tory majority has some considerable advantages for us, especially as Farage has quickly reversed his resignation and will stay a prominent figure on the British political scene.

What we now envisage is a Tory government doing what it has really wanted to do all along, but it will be increasingly beset with internal party divisions, restless and quite far right-wing backbenchers and all kinds of complications over Europe. Meanwhile, Labour will spend the next four months arguing about the leadership and systematically failing to provide opposition. The Tories will become more and more confident so that when the cracks in the edifice start to appear next winter they will look more and more for enemies to divert attention. We make great enemies!

What is even better is that they are already planning to tighten up on their counter-terror legislation which is of course exactly what we want. Just think – more surveillance, new laws restricting freedom of speech, more arrests, tougher sentencing and more young people jailed for long periods. UK prisons are already hugely useful in proselytising for new recruits and as the numbers rise, so will the opportunities. Moreover, all of this will be in a climate of increased xenophobia and fewer life-chances for our young Muslim brothers and sisters. Remember, we are in this for decades.

Perhaps it will all come apart for Cameron and his ilk as it did twenty years ago for that hapless John Major but I doubt it. In any case, it will not be too quick, so we have at least a couple of years to savour. I have to say that, on reflection, this was an excellent result and far better than we feared. For someone like me watching the British scene it is really good to be alive!

14.7 Sixth Letter (22 July 2015)

The writer continues to show confidence in the future of his movement, pointing to successes, as he sees them, across south Asia, and indicating that the movement has potential in Russia. In assessing the UK, concern that the all-too-peaceful Jeremy Corbyn might be a problem if he ever gets into power is moderated by the belief that the chances of this are minimal.

This is my sixth letter to you since we started our correspondence last October and while I know that you, in the relative safety of western Baghdad, do not share my views, at least our old friendship allows us to continue to write. You ask again after my brother and I remember that when I wrote to you in May he was back in good health, training young recruits and hoping to be sent to Afghanistan as our campaign there expands.

Well, as you might expect from him, the training he was running involved a near total commitment and the result has been that his recovery has been slower than expected. As a result he has been put on lighter duties for the next two months. In one way I am very pleased because while I would have been hugely proud of him had he been sent to this important theatre, at least I now see more of him. He told me last week during Eid that his next task will most likely not be a posting to Afghanistan but to Libya, where our links are strengthening by the day.

From your most recent letter I see that you are still very keen to learn what I think of the progress of our war, and there is much to report. First, I should say that my work is still focused on being an effective part of the SOBRA analysis centre here in Raqqa, and I remain largely concerned with the UK, not least because the post-election environment needs to be followed closely in relation to its implications for our work.

Even so, the UK is not quite as important as it was, and one of my most significant duties is to contribute to a very small cell that seeks to maintain a realistic overview of our overall campaign beyond our established caliphate and, indeed, beyond the Middle East.

I'll begin by giving you our perspective on this vast region, and say that while there are interesting developments in south Asia and sub-Saharan Africa, the main focus is on Afghanistan, Libya and a third country which may surprise you – Russia – which is attracting much interest among our most senior leaders. Before covering Russia, though, I'll just update you on the other two.

Afghanistan continues to show progress, but it is variable, mainly because of our uncertain relationship with the Taliban and some other armed opposition groups. You always have to remember that the Taliban simply do not share our vision of a transnational caliphate ultimately encompassing the whole world. They are essentially Pashtun nationalists with little more than a veneer of religious purity. As such, they are all too ready to compromise with local norms in the pursuit of power, an attitude that we find utterly frustrating.

Even so, we have been able to make some strong links with some of them and have even been part of significant campaigns, often bringing our paramilitary expertise to bear in a manner that the more prescient and intelligent of them welcome. Indeed, what we have recently heard is that the crusader leadership in Afghanistan think we are such a serious threat that they may reconsider their original plan to withdraw their remaining 9,800 troops by the end of next year. As one of their generals put it, they think we have moved from a 'nascent' threat to one that is "probably operationally emergent".

What a weird phraseology, yet it is music to our ears! If the crusaders still have thousands of troops in the country when they elect their new leader next year, then we expect that whoever gets to their "White House" will be more militaristic than Obama. This means that we will look confidently to many more years of war and thousands more recruits to our cause.

In Libya, our progress is steady and has none of the problems of the narrow nationalism of the Pashtuns in Afghanistan. Instead, though, it is a country that is so unstable and disjointed that any effective organisation can make rapid progress. The crusaders did us an immense favour in getting rid of Gaddafi – even now they fail to recognise how brutal and effective he was in suppressing our cause.

One effect of that was that thousands of young Libyans left the country and came to join the fighting in Iraq after 2003, and many of those that survived are moving back to Libya to aid our programme of expansion. They are a powerful force – highly experienced paramilitaries including some who fought JSOC in 2004–07 but also having intimate knowledge of Libyan society. All in all we see Libya as an environment of great promise.

Then there is the real surprise – Russia. You will know that the Chechens and others from the Caucasus have long been important to our cause but what is striking is how many of them within Russia share our aim of a Caliphate in all its detail. I do not know the exact figures but I understand that we now have close to 2,000 Russians fighting for us here, with tens of thousands more pledging allegiance within their own country.

What we find hugely welcome is that their president, Putin, knows no other way but to crush dissent with persistent force. He has no idea at all of how counterproductive this will be in the long term. After all, Russia has at least 16 million Muslims, more than a tenth of the population, and his actions against some of them cause anger among far more. This in turn helps stir up bitter Islamophobia among the majority population, making even more people uneasy and having the potential to provide thousands more recruits to our cause. He really is clueless!

But let me move on to my own area of specialism, Britain, which occupied so much space in my last two letters. You will remember that back in March what we would really have liked from their general election was a Conservative-UKIP coalition, ideally with Farage as deputy prime minister and interior minister (they call the post "home secretary").

Just before the election we had had sleepless nights that Miliband might get in, ending up with a government even more cautious than Obama. In the event, things have turned out rather well with a Conservative government using the chaos in the Labour Party to force through many policies. In my last letter, I predicted that "Labour will spend the next four months arguing about the leadership and systematically failing to provide opposition". How prescient I was!

Yet there is just a bit of a worry, because the one candidate who is bitterly opposed to doing what we want – expanding the war – is currently doing rather well. Indeed Mr. Corbyn might even get elected and while we think that will bring on years of infighting leaving the Conservatives to do what they want, we have this niggling worry that any discussion involving any kind of opposition to the war is not good news for us.

I personally also have a slight concern that Cameron's speech earlier this week about tackling what he calls extremism was not as hard line as we had hoped, in spite of the deliberate provocation by our Tunisian associates.

At least he only covered the domestic front and it is clear that he wants to extend the British war to Syria later in the year. For us, this is far more important.

There is much more I could write about, especially in relation to our recent progress in Syria, but I have work to do now and will have to leave that to another time. Be assured, though, that we are making considerable progress so do not believe the crusader propaganda. Remember that they were saying we would be finished by the end of last year. Instead we are stronger, with hundreds of recruits joining us every week.

When Britain joins the war in Syria we are fully confident that it will help to get even more from Britain. We might have to do some more provoking but I am confident that our leaders are following events closely and will decide if and when to order it.

14.8 Seventh Letter (15 October 2015)

Three months further on and ISIS is experiencing the full brunt of the air assault. He continues to present a positive slant but there is an element of concern in the language. The confidence in ultimate success perhaps dented by the losses, though he seeks comfort from the possible effects of those very losses in terms of a positive impact on recruitment and the progress that he sees ISIS making in the wider world, notably Afghanistan and Libya. Russian involvement in the war in Syria is an asset, allowing Moscow to be framed as anti-Islamic.

Thank you for asking after my brother. In my last letter I told you that he was recovering from his injuries but that it was turning out to be a longer process and I thought it likely that his next posting would be to Libya rather than Afghanistan. That turned out to be the case and he left for Libya two weeks ago. I have not heard from him since which is worrying. But my other sources say that none of our recent deployments have so far been engaged in heavy fighting, as our associates in Libya are making very good progress.

You asked me to give my view on the progress of our war, and I hope our long-term friendship can survive the obvious differences that we have. Since it is three months since that letter and much has happened, I will try and tell you what the thinking is here in Raqqa.

As you know it is now a year since I first wrote to you and that makes it three years since I travelled here to join this historic cause. I still wish desperately that I could join in the fighting, but with these injuries it is just not possible as I would only be a burden on my friends. We do have quite a competent prosthetics department at the hospital here in Raqqa but they are very hard pressed and are concentrating on lower-limb losses. Compared with the injuries some of our fighters are suffering my missing arm is insignificant.

Instead I am continuing my time in Raqqa with my analytical work for SOBRA although my responsibilities have changed with time. You will recall that at the time of the British general election I had been tasked with providing analysis of the implications of the possible outcomes. Our hope was that a weak Conservative government would be propped up by an expanding UKIP, ensuring a properly right-wing government committed to war in Syria and Iraq and enacting hardline policies towards immigrants in the UK. Such a result would have been just right for us but things did not really turn out as we had wished.

Even so, Cameron got back in and is reasonably secure, so that was good news for us. I then reported our growing concern about what was happening in the Labour Party's frenetic leadership campaign. In the event our fears were justified because they have ended up with a leader, this Mr. Corbyn, who is essentially anti-war. This is quite a serious issue for us, since our leaders believe that there is considerable scope for increasing recruitment from Britain, provided it gets fully involved in the air-war here in Syria.

Fortunately for us, Corbyn has opposition within his party and only yesterday a number of his members of parliament defied him over an economic issue. We are confident that sufficient numbers will be misguided enough to back Cameron if and when it comes to a vote on bombing Syria.

Beyond that, there is much to report, not least because I have returned for most of the time to my earlier work – analysing overall western attitudes to our mission. In Afghanistan, our progress has been far better than even the most optimistic of our leaders expected, and our position is greatly helped by the expansion of the Taliban. This is not so much because we work with them – in many cases we take over from them – but because their progress elsewhere ties down so much of the Afghan national army.

The ANA, in any case, is in disarray and this is hardly surprising since they have lost over 5,000 of their people so far this year, leading to rampant desertions and singularly low morale. The Americans are now planning to keep their 10,000-plus troops there until after the end of 2016 – and since whoever replaces Obama is likely to be more hawkish, that is good news for us. It's one more example of a Crusader occupation for us to exploit.

In Iraq and Syria, the last few weeks have seen rather desperate attempts by the Americans and the other Crusader forces to paint a picture of success. These include the very recent American claim that they have killed 20,000 of our people in the past fifteen months, 5,000 more than they were reporting in July. They also say, extraordinarily, that they are not killing civilians. The reality is far different and we are able to communicate the many hundreds of personal stories every month, thanks to our remarkable and much-expanded media-outreach system.

What the Americans and their coalition do not realise is that though our losses are considerable there are three elements that are working against them, rooted in our long-term aim of persistently showing that we are the true protectors of Islam under Crusader/Zionist attack.

The first is that every one of our fighters and civilians killed will have scores of family members and friends, most of whom will be bitter at their loss and determined to avenge their killing.

The second is that our losses are more than made up with new recruits from across the world. Even the Americans have accepted that around 30,000 people have joined the cause from abroad, twice the size of their estimate barely a year ago.

The third is that our core paramilitary leaders are simply the best in the world – hardly surprising since they got their combat-training against the special forces of JSOC in Iraq barely a decade ago.

You ask about our attitude to the Russian intervention and I have to say we welcome it with open arms. It is actually a small-scale operation that will not be sustainable for any length of time since all the resupply has to come either by sea from Crimea, which is a ten-day round trip, or by air transport via the convoluted route over Iran and Iraq.

What the Russians are very good at, though, is making much political capital out of very little, at the expense of the Americans. Take their much-vaunted cruise-missile strikes that I saw got wall-to-wall coverage in the western media. The Russians fired twenty-six missiles, and seemed to get away with presenting these as unique weapons – whereas the Americans started firing their own cruise missiles at Iraq in the first war 24 years ago!

Moreover, the Russian missiles had a failure rate of 15 percent, yet barely two weeks earlier the Americans had fired 47 cruise missiles at Syrian targets with a failure rate of under five percent. This was simply not reported in the general western press, though our media department was quick to monitor it.

More generally, the Russians want to prop up Assad and that hardly bothers us because our engagements with the Syrian army are so few and far between. Also, if Russia supports Assad this really angers the Turks, with the result that they will go even easier on us than they have been doing over the past year.

Where Russia really aids us, though, is in Russia itself, with all the added recruitment potential among more than 16 million Russian Muslims. So far about 2,000 have joined our cause here in the Caliphate and we also have links with powerful paramilitary forces within Russia, especially in the Caucasus but even in Moscow itself.

Now that Russian strike-aircraft and helicopters are killing Muslims here we can publicise that massively through our networks (it does not matter that the Muslims being killed may not be linked to us). I can tell you that at the first sign of Russia moving military forces to Syria our SOBRA chiefs sent out an urgent message throughout the Caliphate to identify Russian speakers. We already have a taskforce of 20 including many dialect specialists, and they are working night and day to communicate with potential recruits. We expect a bonanza.

You might also like to know that our leadership is already prepared for Britain to start bombing Syria and I am proud to say I will be heavily involved. What has been done is that a number of fluent English speakers, all graduates, have been identified and given preliminary training in script-writing and presentation. They will all be ready to boost the existing unit, which I will head, as soon as the British parliament votes to bomb.

As you can imagine these are very exciting times and I have come to feel that I can make a difference. At long last I will be able to aid the cause while also avenging all the members of my family killed by the Crusaders in Iraq. It has been a long time coming but now I can begin to feel proud – it is all so clearly worthwhile.

14.9 Eighth Letter (4 February 2016)

By now the writer's work has been redirected to covering the American presidential election campaign as it takes shape. His assessment includes the then forlorn hope of what, from ISIS's perspective, would be the ideal result: a victory for Donald Trump! He is hopeful, though, that any result will deliver a more hawkish administration than Barack Obama, which would be welcome among the leaders in Raqqa.

Thank you for your letter and all the news from Baghdad. I am glad to know that your sister has recovered from her injuries, especially as medicines are in such short supply. I hear that the collapse in oil prices is already leading to a surge in the black market for medical supplies. Is that true?

Thanks also for asking once again after my brother. I am pleased to say that I have just heard from him at the end of a hugely worrying time since he left for Libya at the end of September. He tells me that he is fit and well and very heavily involved in the forays from Sirte down to the oilfields. As we had heard, our militias in Sirte have expanded their hold, having been hugely helped by nearly 2,000 of our fighters who have gone to join them.

What few people in the western military realise is that the great majority of them are actually Libyans returning home, having fought so well against the crusaders and apostates in Iraq and Syria. They are therefore dedicated to extending the Caliphate into their own home territory and also exceptionally experienced in combat. My brother tells me that the crusader military are increasing their use of drones and also using special forces raids but to little effect, so we expect to see systematic bombing raids starting soon, with the inevitable civilian casualties in turn leading to greater support for our cause.

You ask me what I am now doing – do I still work for SOBRA and indirectly undertake analysis for the leadership, and does this still relate to my British experience? In one way I am still frustrated that I cannot be on the frontline, but I am slowly accepting that this is a necessary consequence of losing my arm in that raid. At least it was my left arm. As I said in my last letter we have a good prosthetics team here in Raqqa but the priority has to be lower limb replacements and I am now told that it will be some months before I can get a prosthetic arm.

At the same time my role has changed and in quite an exciting direction. You will recall that my analytical work around the time of the British general election was in assessing how the British might react in a new parliament. We had been worried that Miliband might get in whereas our greatest wish was for a minority Conservative government having to go into coalition with UKIP. Our vision of Cameron beholden to Farage and having to make him home secretary and deputy prime minister would have been like heaven to us!

It wasn't to be but at least Miliband failed, so our work on UK political analysis was downgraded. Even so, I am tasked with spending a small part of my time keeping a watching brief and have to say that the Corbyn situation is a bit of a worry. Although nearly all the British commentators think he will fail as Labour leader we have a sneaking suspicion that he is striking a chord with more people than they realise, with this including offering a voice to perhaps 20 million people who do not agree with Cameron's stance on the war. It would be a disaster for us and the movement if he was ever to get into Downing Street, but at least that is some years off.

Because of this I have been asked to run a small but highly knowledgeable team analysing and advising on US politics throughout this election year. You may remember that I spent parts of two years in Washington so I find it particularly thrilling to be asked to analyse the race for the White House, given that the outcome will be so crucial to our future.

As far as the contenders are concerned, what we would like most would obviously be a Trump victory – even better than having Farage sharing power with Cameron in London! Our nightmare, of course, would be a Sanders victory but in spite of the Iowa result we do not currently expect him to get the Democrat ticket.

In addition to our analysis of open sources, we do have some insider voices reporting to us, and also have a wider circle of contacts who add a more general perception. The consensus view is that Sanders has some way to go but the main impact will be to make Clinton appear more progressive than she really is, but only for the duration of the primaries. The odds are still on Cruz for the Republicans but my own view is that his support will recede and that it will be Marco Rubio who comes to the fore. We will then have a Clinton-Rubio contest and we will be reasonably content with that.

You may ask why we would be content with such a contest but I think you have to look at the bigger picture. America is moving slowly but surely to the right, and even Obama is steadily allowing a build-up of US operations against us and our associates. This is just what we want as the true defenders of Islam, and since Clinton or Rubio will be more hard line than Obama, we will have more defending to do, which works wonders for recruiting more followers.

In spite of our recent minor reverses our leadership is in good heart and are buoyed by three positive developments. The first is the American move to the right which is well-nigh certain to continue and the second is our programme of overseas attacks designed principally to stir up as much Islamophobia as possible. More attacks are being planned for western Europe but I understand that an even greater priority is Russia. The possibility of really exacerbating anti-Muslim feelings towards Russia's 16-million strong Islamic minority is a prize really worth aiming for.

It is the third development, though, that is the hugely welcome surprise – the crisis over refugee flows to Europe. I don't think that any of their politicians, even the few intelligent ones, have any idea what is being stirred up. Can they not see the impact of desperate people, including young children, facing razor wire and being violently repelled by armed riot police as they try to get into Europe? Can they not see that to Muslims across the world, let alone the Middle East, the vision is just proof positive that Europeans hate and despise them? This hatred will last for years and years.

So put it together – America goes more hardline, the wars intensify, the refugee flows grow, Europe turns its back as anti-Muslim feelings increase, community disorder and violence become the order of the day. The end result? Many thousands more recruits to our cause.

Perhaps you can understand why someone like me is quietly optimistic. Never forget, we are fighting a cause that may take a century or more to achieve. Our opponents, the 'far enemy' really do not have a clue, and long may that last!

14.10 Ninth Letter (28 July 2016)

The war is hardly going well for ISIS, given the intensity of the air attacks. The writer, however, says little about this. His delight at the Brexit result is palpable, seeing it as a major part of the move towards xenophobia and Islamophobia in western Europe, trends likely to make Muslims more fearful and at least some of them likely to see ISIS as the true way forward. He rejoices yet again that the west has so little understanding about the nature of ISIS and the new Caliphate. From his perspective, ISIS wants to be attacked and seeks to damage community relations in Europe as much as possible, thus damaging society from within. For him, plenty of politicians and much of the western media are giving ISIS precisely what it wants.

Thank you for your letter and I am pleased to hear you were nowhere near central Baghdad when the bomb went off. I know we have had our differences, but do believe me when I say that war has consequences which even those involved can regret.

I must apologise for taking nearly six months since my last letter but things have been very hectic here as the air strikes – Russian as well as American – have come closer to the city.

At least I have now been fitted with a prosthetic arm and have been learning how to manage with it. Since I lost my arm in that American raid several years ago, I had got used to managing without, so the new arm is taking some getting used to. It was amazing to get it though, given the pressures that our hospitals are under, and while it is hardly state of the art it is certainly making a difference.

You ask again after my brother. I have some really surprising news of him. You will remember that back in February I had just heard from him in Libya, even though I knew he had been there for some months. Well I have just heard through a friend that he is now in Bangladesh! Not only that but he is part of the small team that has been involved in the rapid development of our organisation there, including the action against the foreigners a couple of weeks ago.

I understand that the operation did not go quite according to plan as they had expected to find more Americans there rather than Italians, but the effect on Bangladeshi politics has still been considerable. It has also demonstrated categorically that ISIS has 'reach'. It also coincides with our action in Kabul, which gives the lie to Ghani's claim that ISIS in Afghanistan is finished.

You ask about my own work. It has mainly been about continuing to analyse and assess the US presidential election for our SOBRA intelligence centre. But I also still continue with my watching brief on the UK. You will remember that this was my main role around last year's election there, and that we had been concerned that Miliband might win and bring in a less antagonistic attitude towards us. That was not, of course, what we wanted. The more the Crusaders attack us the more support we can generate.

In the event the result was fine for us. We had this dream of a weak Conservative/UKIP coalition with Farage as Cameron's home secretary, but we knew full well that it was unlikely. We were happy to settle for a rather unstable Conservative majority. Corbyn coming in for Labour was, as I said in an earlier letter, a bit of a worry since the last thing we wanted in UK politics was an out-and-out peacenik, but we had few concerns that he would ever get there given the quintessentially right-wing nature of the British press.

Where Cameron turned out to be an utter gift for us was his need to assuage his right-wing with a referendum on the EU. You can imagine our utter delight at the outcome – roll on Brexit and all the upheaval! All we want now is Marine Le Pen winning the French presidency next spring, followed by an even messier Frexit and the break-up of the whole Crusader edifice! But more of that later.

As I said, my current job is with the US election and I note that my last letter to you mentioned rather wistfully the possibility of Donald Trump getting the Republican ticket. It seemed frankly unlikely back in February, but just look at what has happened since! Not only has he got it but he looks like he is levelling with Hillary Clinton as polling day approaches. Our senior planners are currently working hard to decide how we might intervene to improve his chances and I am pretty sure there will be some interesting developments there.

What our leaders want is a convincing win for Trump and the enacting of his harder-line policies on immigration, religious freedom and the like, coupled with a reversal of his current rather curious brand of isolationism. Expect some anti-American spectaculars both before the election and even more so afterwards if he does make it to the White House. These will be designed to ensure a renewed onslaught on us by the Americans and their European Crusader allies.

You ask me to be frank with you about the current state of the war and I will oblige. You may be surprised to hear how optimistic I remain in spite of the intensity of the war being waged against us. It is true that we have lost many thousands of our young fighters as well as hundreds of women and children, and the recent acceleration in the intensity of the air raids has taken its toll. We have also lost valuable territory in Iraq, especially the fall of Fallujah.

So you might think it reasonable to ask why I remain optimistic.

Look at it through other eyes, though. In Iraq the apostate government continues to favour the Shi'a and marginalises our natural supporters. Some of the behaviour is grotesque, as with the abandoning of tens of thousands of refugees from Fallujah even though they were refugees only because of the government assault on their city. Then there is the overwhelming presence of the Shi'a militias without which the government could not maintain control.

Put these together and are you surprised that support for our cause is rising in Iraq at the very time that our prospects are apparently so dim? Furthermore, our military planners are already preparing and positioning for a long-lasting insurgency against the Americans, Iranians, Shi'a militias and the Iraqi government. In doing so they are sure of plenty of support from many wealthy backers in the Gulf who are horrified at the rise of Iran and see the Shi'a crescent from the Mediterranean to the Arabian Sea as evolving before their eyes.

I also suspect that western analysts have no idea about the impact of our round-the-clock social media reporting of the Crusader air war and of the families torn apart by the bombing and drone strikes. These reports lead to palpable anger across the world and serve most effectively to aid our recruiting base.

There is a further element, however, which helps explain my positive outlook. Let me explain. If we go back two years, just after the startling successes in Iraq, all our emphasis was about the actual achievement of a new Caliphate out of what was previously northern Iraq and northern Syria. That was lauded as the way forward, in marked contrast to the failed policies of al-Qaida a decade earlier. But our top planners were already looking beyond this.

They correctly saw that the Crusaders would come at us with massive force and might eventually cause us huge problems, even to the extent of taking parts of the new Caliphate away from us. In these circumstances they reasoned that even the temporary existence of the Caliphate would be of huge symbolic importance, and its crushing by the Crusaders would be a call to arms in the coming years.

But they went further, in determining to take the war to the Crusaders. By the start of last year we were developing a range of capabilities. Some of these were under our direct control, others subject to strong influence and some technical support, and still others were no more than inspired by our work. A few of the recent attacks were completely unrelated, even if we moved quickly to claim them.

The effect of all of this, especially in France and Germany, has been to increase the sense of insecurity, accelerate the rise of anti-Muslim bigotry and increased support for extreme right-wing parties. We are most interested in Germany and France, but work is also under way to enhance our support-base in several other countries, including the Netherlands, Austria and Britain.

Even so, our main target has to be France and we are growing increasingly confident of influencing next spring's presidential election in favour of Marine Le Pen. We may not have got Farage into the British government but Trump in the White House and Marine Le Pen as president of France will make us very happy indeed.

I have to say that even though I worked for three years in the United States and studied for a year in London, I am still amazed that senior policy people and analysts in both countries seem so unable to understand ISIS and its aims. Even after 15 years of war, the Crusader states are no more secure than they were. Yet they do not even try to learn why. For the sake of our mission, let us hope they never do.

14.11 Tenth Letter (8 December 2016)

The tenth and most recent of the letters. Here he acknowledges that the days of a defined geographical Caliphate might be coming to an end. Yet he does not see this as a disaster, given the election of Trump and the renewed xenophobic tendencies in Europe. Indeed, he now sees the Caliphate as a symbol for the long term and is deeply frustrated that he is not allowed to fight. He hopes yet to do so, concluding that this might be his final letter.

Thank you for your letter and for asking after my brother. I must apologise for taking so long to reply, especially as I last wrote to you nearly five months ago. There is much to say. But first, what of my brother? Last time I wrote he had just gone to Bangladesh to help the establishment of our new cell there. In the event he stayed a month and then went to Libya, much as I had suspected he would.

As you will have heard, the apostate militias and the crusader-backed government started a concerted attack on our movement last May, expecting to take our main centre of Sirte in a matter of weeks. Three months later and in the face of failure, the Americans were forced to bring in their aircraft-carrier and start air-strikes, but even with intense bombardment our forces held out until this week.

My brother's role is to help organise the next stage. Libya is in utter chaos and many of the factions share our vision, so we see no difficulty at all in regrouping – indeed, many of our fighters have already melted away from Sirte to reform units elsewhere. There is much to play for, especially as across the border in Egypt where al-Sisi is so intent on repressing our cause that he is playing right into our hands. Already our movement is thoroughly embedded in like-minded groups in Sinai, but what is far less known is the growing presence of our supporters in Cairo and other cities.

You also ask after me, and I have to report both good and bad news. The good news is that my new arm is functioning well, and I am able to continue my analytical work with SOBRA with even greater efficiency. The bad news is that my superiors absolutely refuse to let me join our fighters in Mosul, even though I am fully fit and able to defend with the best of them.

My superiors tell me bluntly that the analytical work I do for them on political developments in Britain, Europe and the United States is far more important. I know that my three years in London and Washington in the quite recent past do give me an understanding of what is happening within the far enemy but I often wish that I had never gone there. That way I could have already fought much more directly for the cause.

You and I have been friends for many years and that friendship has held even though you are in Baghdad and I am in Raqqa at the heart of the Caliphate, and we take such radically different views on the future path of our faith. I have to say that my superiors would be deeply antagonistic if they knew I continued with this correspondence, but for now I want to do so, although you will have gathered that there is another motive in that.

As you will know, my work involves accessing a very wide range of sources from the enemy including press, radio, TV, popular culture, military journals, the web and social media. But I also know that you do sometimes pass on my letters through your own social media. In its way that is helpful since I can learn how people elsewhere respond to my comments, in turn enabling me to understand better the extraordinary misinterpretations that are so common across the western world.

In your last letter you asked me to explain how morale within the movement could possibly be maintained when facing retreat after retreat – the loss of Fallujah and Ramadi for example, as well as the impending loss of Mosul. Indeed you report the view that our Caliphate will be destroyed by the end of next year, thus ending our extraordinary journey. You will not be surprised if I beg to differ, indeed that I get exasperated with such views, but I will try and respond in a rational way. After all, my post in SOBRA is predicated on just such rational analysis!

According to sources from the far enemy, primarily the Pentagon, since their air war started over two years ago they have attacked us with over 16,500 air raids and dropped over 60,000 bombs and missiles hitting over 30,000 targets. This is against our forces that have usually been estimated by them to number 25,000–30,000. To put it bluntly, they have used enough force to hit every one of us with two bombs or missiles! Indeed the Pentagon claimed back in August that they had killed well over 40,000 of us.

Something isn't quite right here, is it? Either they are lying about killing us or we had far more people ready to fight or tens of thousands have joined us since the start. The reality is that their air war has indeed been extremely intense and they have killed very many thousands of our fighters and thousands of civilians but we had many more when the war started and many thousands have joined our cause since.

Look at what is really happening in Mosul. When their assault started seven weeks ago they had at least 80,000 attackers from half a dozen countries backed up by plenty of air power including drones, all determined to wipe out our few thousand defenders. In an absolute blaze of publicity they reported rapid progress into the city within two weeks.

Then what? It all went quiet and now there is scarcely any reporting from their side. If you search through their news sources you will find buried away the fact that they have been fought to a standstill and have taken control of less than a third of the eastern half of the city in seven weeks. Moreover, our main forces are still west of the Tigris. You can even find some Iraqi security sources admitting that they could still be fighting us in Mosul next summer!

Even so, I will say to you privately that we are experiencing a very heavy onslaught, one that has been going on for more than two years. I will also accept that the Caliphate as we have established it may not last for ever and will, instead, be part of a much longer process that will stretch over decades.

But before I come to the significance of that, let me just remind you of some of our spectacular achievements of recent years and indeed months.

For a start, right across the world they have been forced to spend trillions of dollars since 9/11 on their failed wars as well as vast sums of money on their own domestic security. Look at the security precautions at every airport, every legislature and every government office and look at the huge expansion of security forces and intelligence agencies and the across-the-board increases in surveillance. All of this to try and defeat a far, far smaller number of determined opponents who have little more than light arms and improvised weapons to fight against the best trained and best equipped military in the world.

Then look even more at the extraordinary political changes. In Britain there is Brexit, won partly by scare tactics. Remember the "breaking point" poster of thousands of desperate refugees actually presenting an existential threat to the UK. From our perspective we could hardly have asked for more!

Look at the rise in Islamophobia across Europe, at Geert Wilders in the Netherlands, Marine Le Pen in France, the far-right government in Hungary, the almost-success for the far-right candidate in Austria. All of this has been very largely down to us and people like us. Don't you see why we are so sure that we will eventually win?

I mentioned Libya earlier, but look also at Afghanistan where even the Americans are hugely worried about the insecurity. They say openly that the government only controls 60% of the country, and I can assure you that we are increasingly active there.

Remember what I said to you in an earlier letter when I was working on the analysis of the US election:

> As far as the contenders are concerned, what we would like most would obviously be a Trump victory - even better than having Farage sharing power with Cameron in London! Our nightmare, of course, would be a Sanders victory but in spite of the Iowa result we do not currently expect him to get the Democrat ticket…
> So put it together - America goes more hard line, the wars intensify, the refugee flows grow, Europe turns its back as anti-Muslim feelings increase, and community disorder and violence become the order of the day. The end result? Many thousands more recruits to our cause.
> Perhaps you can understand why someone like me is quietly optimistic. Never forget, we are fighting a cause that may take a century or more to achieve. Our opponents, the "far enemy", really do not have a clue, and long may that last!

Since then, not only has Mr Trump won but he has already appointed two really hawkish retired generals, one to head the Pentagon and one to be head of his national-security office – two men who have been particularly anti-Islamic – Perhaps you can understand why I am so optimistic, even if deeply frustrated that I am not allowed to fight on the frontline.

But let me finish on one more point – what in the West they might call a "worst-case scenario." As I read the western media and assess their changing attitudes, one thing that comes across is their belief that the Caliphate is finished. They do expect Mosul to fall to them eventually, although even that is no longer certain. Even so, they believe it will happen and they expect the same to apply to Raqqa. From their perspective that will be the end of it all.

I very much doubt that our Caliphate will be ended any time soon. But if that did happen in the next two or three years then what they miss entirely is that the very existence of an actual Caliphate, even if it lasted just five years or so, is what will count. It will have served as an extraordinary symbol of what can be achieved by a small but utterly determined group of the true faithful when pitted against the world's most powerful military forces.

We do something that they simply cannot even begin to comprehend. Our aim, rooted in our faith, is to bring about a global transformation to true belief and we as individuals are just one small part of a process that stretches far beyond our own lives. We are, in a real sense, engaged in an eternal struggle and they cannot understand this, which is why we will win.

This may be my last contact with you in this life, as I hope very much to be released from my post very soon to join that fight as I have wanted to do for so long.

Part VI
Selected Texts: Putting It Together

Chapter 15
Peace Research, Then and Now (1999)

Paul Rogers and Oliver Ramsbotham

15.1 Introduction

Despite ancient origins, it was in the years after World War II that peace research, as a formal field of study with its own academic institutions and professional journals, was established.[1] The failure of the various peace, socialist and liberal internationalist movements to prevent the outbreak of World War I had inspired renewed efforts thereafter, particularly in Europe and the USA, to develop a more thoroughly grounded 'science' of peace in order to learn how to prevent future wars.[2] But, as van den Dungen comments, the many proposals and initiatives at this time were isolated and individualistic, in which 'exhortations far outnumbered realisations'.[3] The future field of peace research would be able to draw on empirical evidence gathered and analysed in the interwar years by scholars such as Pitrim Sorokin, Lewis Richardson and Quincy Wright,[4] and on other pioneering work in

[1]This text was first published as: Paul Rogers and Oliver Ramsbotham, "Then and Now: Peace Research—Past and Future", *Political Studies* (1999), XLVII, pp. 740–754. The permission to republish this text was granted on 2 March 2018 by C. Myles on behalf of SAGE.
[2]See P. van den Dungen, 'Initiatives for the pursuit and institutionalisation of peace research in Europe during the inter-war period (1919–1939)', in L. Broadhead (ed.), *Issues in Peace Research* (Bradford, University of Bradford Press, 1996), pp. 14–32.
[3]van den Dungen, 'Initiatives for the pursuit and institutionalisation of peace research', p. 27.
[4]P. Sorokin, *Social and Cultural Dynamics* (New York, American Books, 1937); L. Richardson, *Statistics of Deadly Quarrels* (Pittsburg, Boxwood, 1960) (published posthumously); Q. Wright, *A Study of War* (Chicago, University of Chicago Press, 1942).

P. Rogers, *Paul Rogers: A Pioneer in Critical Security Analysis and Public Engagement*, Pioneers in Arts, Humanities, Science, Engineering, Practice 21,
https://doi.org/10.1007/978-3-319-95150-8_15

as yet unintegrated fields.[5] Nevertheless, the sustained development of peace and conflict research in the form of institutional growth and conceptual definition had to wait until the post-1945 world, when the added threat of nuclear weapons created a new urgency.

Before outlining the subsequent historical evolution of peace research,[6] we suggest that the following seven features are characteristic of its development and mark it out as a defined field of study:

a. A concern to address the root causes of direct violence and to explore ways of overcoming structural inequalities and of promoting equitable and cooperative relations between and within human collectivities. This led to debate between those who espoused a 'minimalist' agenda and those who advocated a 'maximalist' agenda, as commented upon below.
b. A realization that an interdisciplinary response would be required. Given the multi-faceted nature of violent conflict, analysts would need to supplement an international relations approach with insights from the other political and social sciences, as well as from social psychology, anthropology and other disciplines. This would lead to conceptual enrichment, but would cause disputes about appropriate methodologies and theoretical frameworks.
c. A search for peaceful ways to settle disputes and for non-violent transformations of potentially or actually violent situations. This would not mean endorsing the

[5]Such as Mary Parker Follett's innovative 'mutual gains' approach in organisational behaviour and labour-management relations in H. Metcalf and L. Urwick (eds), *Dynamic Administration: the Collected Papers of Mary Parker Follett* (New York, Harper, 1942); Kurt Lewin's work on the social-psychology of group conflict summarized in *Resolving Social Conflicts* (New York, Harper, 1948); Crane Brinton's analysis of revolution in *The Anatomy of Revolution* (New York, Norton, 1938); and various pioneers of international interdependence theory such as Norman Angell, Ramsay Muir, Francis Delaisi, Charles Merriam and David Mitrany, see J. de Wilde, *Saved from Oblivion: Interdependence Theory in the First Half of the 20th Century* (Aldershot, Dartmouth, 1991).

[6]For reviews of the history of peace research see J. Galtung, 'Peace, violence and peace research', *Journal of Peace Research,* 6 (1969), 167–91: D. Dunn, 'Peace research', in T. Taylor (ed.), *Approaches and Theory in International Relations* (London, Longman, 1978), pp. 257–79; J. Galtung, 'Twenty-five years of peace research: ten challenges and some responses', *Journal of Peace Research,* 22, 2 (1985), 141–85; M. Banks, 'Four conceptions of peace', in D. Sandol and I. Sandole-Staroste (ed), *Conflict Management and Problem-Solving: Interpersonal to International Applications* (London, Pinter, 1987); P. Wallensteen (ed), *Peace Research: Achievements and Challenges* (Boulder CO, Westview, 1988); A. Groom, 'Paradigms in conflict: the strategies the conflict researcher and the peace researcher', in J. Burton and F. Dukes (eds), *Conflict: Readings in Management and Resolution* (London, Macmillan, 1990), pp. 71–100; A. Mack, 'Objectives and methods of peace research', in T. Woodhouse (ed), *Peacemaking in a Troubled World* (Oxford, Berg 1991), pp. 73–106; C. Stephenson, *Peace Studies: the Evolution of Peace Research and Peace Education* (Hawaii, Institute of Peace, University of Hawaii, Occasional Paper No. 1, 1990); E. Boulding (ed.), *New Agendas for Peace Research: Conflict and Security Re-examined* (Boulder CO, Lynne Reiner, 1993); J. Balazs and H. Wiberg (eds), *Peace Research for the 1990s* (Budapest, Akademiai Kiado, 1993). R. Ellis and T. Turpin (eds), *Rethinking Peace* (Boulder CO, Lynne Reiner, 1994); See also H. Miall, O. Ramsbotham and T. Wooodhouse, *Contemporary Conflict Resolution,* (Cambridge, Polity, 1999), ch. 2.

status quo, since unjust and oppressive systems were seen as some of the chief sources of violence and war. It did mean the comparative study of peaceful and non-peaceful processes of social and political change; and of ways to prevent the outbreak of violence, or, if it did break out, ways to mitigate it, bring it to an end, and prevent its recurrence thereafter. Within these parameters there was heated debate about the legitimacy and efficacy of the use of force in certain circumstances.

d. The espousal of a multi-level analysis at individual, group, state and interstate levels in an attempt to overcome the institutionalized dichotomy between studies of 'internal' and 'external' dimensions, seen to be inadequate for the analysis of prevailing patterns of conflict.[7]
e. The adoption of a global and multi-cultural approach, which would locate sources of violence globally and regionally as well as locally, and draw on conceptions of peace and non-violent social transformation from all cultures.
f. An understanding that peace research is both an analytic and a normative enterprise. There was to be a major attempt to ground the subject in quantitative research and comparative empirical study, but, in anti-positivistic vein, most of the scholars attracted to the field were drawn by ethical concerns and commitments. Deterministic ideas were rejected, whether in realist or Marxist guises, with large-scale violence and war seen, not as inevitable features of the international system, but as the consequences of human actions and choices.
g. Linked to this was the close relationship between theory and practice in peace research. Most of the founders made a clear distinction between peace research and peace activism (a distinction not always appreciated by critics). Yet nearly all peace researchers insist that theoretical insight must be empirically tested, and many have been more concerned with the policy implications of their research than with its reception among fellow academics.

In short, peace research overlapped with and drew from existing fields of study such as jurisprudence, philosophy, psychology, anthropology, politics and international relations, but was distinct from them in its central concern with issues of peace and conflict, its multi-disciplinarity, its holistic approach combined with quantitative and empirical methodologies, and its normative commitment to the analysis of conditions for non-violent social and political change. Many peace researchers, rejecting realist and Hobbesian conceptions of peace, drew rather on idealist, liberal-internationalist and structuralist traditions, and attempted to develop

[7]For example, in work originating in the late 1970s, Edward Azar, following John Burton, rejected the 'rather rigid dichotomy of internal and external dimensions' in the study of contemporary conflicts, with sociologists, anthropologists and psychologists preoccupied with the former ('civil wars, insurgencies, revolts, coups, protests, riots, revolutions') and international relations scholars with the latter ('interstate wars, crises, invasions, border conflicts, blockades'), *The Management of Protracted Social Conflict: Theory and Cases* (Aldershot, Dartmouth, 1990).

these into a transformationist research agenda.[8] In the words of one of the early pioneers of peace research at the beginning of the twentieth century, the Frenchman Raphael Dubois, 'would it not be wise to endow the science of peace with rich and strong schools, just as one has done for its elder sister, the science of war?'[9] This was the inspiration for what was to follow, perhaps best summed up under the motto: *si vis pacem, para pacem.*

15.2 Peace Research – The Past

Stimulated by the same revulsion against war and mass atrocity that inspired the founding of the United Nations, and further intensified as the nuclear arms race accelerated against the background of deepening Cold War, the late 1940s and 1950s saw the initiation of formal peace research. In Europe, the *Institut Francais de Polemologie* was founded in 1945, and Bert Röling, a judge at the war crimes tribunals in Japan, introduced polemology (conflict research) to the Netherlands during the late 1940s, while Theodore Lenz's Peace Research Laboratory was founded in St. Louis, also in 1945. The concern over nuclear weapons led to what was to be an enduring and significant feature of future peace research – the steady inflow into an essentially social science area of mathematicians and natural scientists, illustrated by the Pugwash movement and the later initiation of the *Bulletin of the Atomic Scientists.*

In the public eye, as in the view of many academics, the late 1950s and early 1960s represented a period of extreme Cold War danger, as the nuclear arms race accelerated and any attempts at arms control, let alone disarmament, were largely side-lined. Public opposition to nuclear weapons reached a climax at the start of the 1960s, and this was also a period which saw some of the early work on non-nuclear defence alternatives. Against this background, the modern era of peace research might be said to date from the establishment of the *Journal of Conflict Resolution* at the University of Michigan in 1957 and Galtung's founding of the forerunner of the Peace Research Institute in Oslo (PRIO) in 1959, but earlier initiatives are relevant, especially in the USA. In April 1951, Kelman and Gladstone published a letter in the *American Psychologist* arguing for serious and systematic consideration of pacifist approaches to foreign policy. This was at the time of the Korean War and led to the publication of the *Bulletin of the Research Exchange on the Prevention of War* in 1952.

Two years later, a group including Kelman worked together at the Center for Advanced Studies in the Behavioural Sciences at Stanford. Along with Kelman

---

[8]See H. Miall, 'What do Peace Studies Contribute Distinctively to the Study of Peace?' paper presented at the Political Studies Association Annual Conference, Nottingham, UK, 24 March 1999.

[9]Quoted in van den Dungen, 'Initiatives for the pursuit and institutionalisation of peace research', p. 14.

were Kenneth and Elise Boulding, Rapoport and Lasswell, a colleague of Quincy Wright. Also there was Richardson's son, Stephen, who had his father's unpublished works on microfilm. The Stanford group gave a considerable stimulus to research on peace and conflict and the Bulletin was shortly afterwards developed into the *Journal of Conflict Resolution,* based at Michigan's newly established Centre for Conflict Resolution. In its first issue, the editors provided two motivations for the new journal:

> The first is that by far the most important practical problem facing the human race today is that of international relations – more specifically, the prevention of global war. The second is that if intellectual progress is to be made in this area, the study of international relations must be made an interdisciplinary enterprise, drawing its discourse from all the social sciences, and even further.[10]

In Europe, PRIO remained part of the Norwegian Institute of Social Research (becoming independent in 1966) and started the publication of the second influential peace research journal, the *Journal of Peace Research* in 1964, three years after Röling founded the Polemological Institute at the University of Groningen in the Netherlands.

If the centres at Michigan, Groningen and Oslo (of which only Oslo now survives) were the main foci of peace and conflict research in North America and Europe in the late 1950s and early 1960s, national and international organizations were also developing elsewhere. They included the Japanese Peace Research group in 1964, the Canadian Peace Research and Education Association in 1966 and the US Conference on Peace Research and History in 1963, which later established one of the other key journals – *Peace and Change.* In 1966 the Stockholm International Peace Research Institute (SIPRI) was established to commemorate Sweden's 150 years of peace, going on to acquire a formidable reputation for its work on issues of armaments, arms control and disarmament and maintaining this focus for more than 30 years, unlike many other peace research centres. In addition to its concerns over nuclear issues, its long-term commitment to work on the control of the arms trade, and of chemical and biological weapons has proved to have remarkable salience in the post-Cold War world. In 1969 the Tampere Peace Research Institute was formed in Finland. Finally, and following a Quaker-sponsored meeting at Clarens in Switzerland in 1963, the International Peace Research Association was established, based initially at Groningen and holding meetings every two years.

The focusing of disciplines not traditional to international relations such as psychology, economics and mathematics on problems of conflict, especially those relating to the Cold War, provided a springboard for vigorous critiques of the prevailing realism in international relations, which were as vigorously rebutted. As Mack has argued, the peace researchers considered traditionalists within IR to be methodologically unsound and prone to unproven assumptions, whereas the latter regarded peace researchers with irritation, seeing them as one part of the

[10] *Journal of Conflict Resolution,* 1, 1 (1957), p. 3.

behaviouralist revolution which claimed 'scientific' understandings of international behaviour which traditionalists considered to be simplistic if not naive.[11]

These differences can be illustrated by contrasting the attitudes of realist IR scholars and peace researchers to East/West confrontation. With the failure of the pre-war League of Nations, and the evident rivalry between western liberal democracies and the totalitarian regimes of the Soviet bloc, realists saw an urgent need to focus on studying the means to maintain the former. A number of peace researchers, on the other hand, along with 'idealists' within IR, saw a need to transcend the western ethnocentric view of the Cold War divide, recognizing what they considered to be legitimate Soviet security concerns and seeing two ideologies locked into a single dynamic of military confrontation and escalation.[12]

Turning to the middle and late 1960s, on the international scene the Cuba Missile Crisis in 1962 perhaps represented the peak of the Cold War danger, and it was followed by a period of quasi-detente which included the negotiation of some significant treaties. The 'Kennedy experiment' of reciprocated confidence building measures resulted in the Limited Test Ban Treaty in 1963, and the early Strategic Arms Limitation Talks got under way. Within five years, significant agreements, including the Non-Proliferation Treaty, had been negotiated, and there was a widespread view that the worst of the Cold War dangers were in retreat.

Parallel with this came the first major controversy within peace research as the original 'minimalist' agenda of preventing war, and in particular nuclear war, as advocated by what might be called the North American pragmatists, was challenged by the broader 'maximalist' agenda insisted upon by European structuralists. Galtung was particularly significant among the latter as he argued that the condition of peace required not just the absence of overt violence but also of structural violence – the effects of economic and social exploitation.[13] The disagreement was at times hard-hitting. For example, Herman Schmid argued that much of peace research was not critically engaged in defining peaceful societies as entities in which justice genuinely prevailed.[14] In this view, which resonated with the earlier positive/negative peace debate, an absence of war could obscure deep injustices which made a mockery of notions of peace. On the other hand, minimalists found the constant expansion of the peace research agenda 'as acquiring the qualities of an

[11]A. Mack, 'Objectives and methods of peace research', pp. 73–6.

[12]More precisely, the 'orthodox' western view that attributed the Cold War to Soviet aggression was countered by at least four alternatives: a 'revisionist' view that attributed it to the global ambitions of capitalist imperialism; a 'neo-realist' view that interpreted it as the great power rivalry to be expected in a bi-polar world; a 'neo-liberal' view that saw it as a dangerous dynamic generated by mutual worst case security preoccupations; and a 'radical' view that saw it as an 'imaginary' war generated by elites on both sides in order to maintain domestic control within their own blocs. For the latter, see M. Kaldor, *The Imaginary War: Understanding the East-West Conflict* (London, Blackwell, 1991).

[13]J. Galtung, 'Violence, peace and peace research', *Journal of Peace Research,* 3 (1969), 167–92.

[14]H. Schmid, 'Politics and peace research', *Journal of Peace Research,* 3 (1968), 217–32.

intellectual black hole wherein something vital, a praxeological edge or purpose, is lost'.[15] This was a criticism made, among others, by Kenneth Boulding.[16]

This dispute was never formally resolved, but most peace researchers came to accept that, in addition to the original aim of learning how best to prevent nuclear war, two other major themes were also now of legitimate concern for those working in the field: North-South global disparities, and what was perceived to be a looming environmental crisis for humankind.

From 1955 through to 1965 it was increasingly argued that the political independence being achieved through decolonization was not being accompanied by economic liberation. For southern states, the global economy was seen as essentially a colonial creation assigning them a subservient role as producers of primary commodities, made worse by numerous trade barriers imposed by the industrialised North.[17] These ideas greatly influenced the early work and aspirations of the Group of 77 and the UN Conference on Trade and Development, and they fitted in markedly with the thinking of peace researchers developing the concept of structural violence. Whereas there was substantial commonality with development studies, this phase of peace research had little in common with developments in international relations, where concern with the study of North-South relations was rare in the extreme.

Peace researchers were also among the earlier social scientists to recognize the significance of environmental issues which had come to the fore in the 1960s, primarily as a result of ecological and toxicological investigations into the impact of industrial societies on the natural environment. Early concerns over pesticides developed into much wider critiques of the impact of unbridled economic growth through pollution, land dereliction and resource depletion. Much of this concern was criticized as introspection by environmentalists lacking concern for the much wider issues of global underdevelopment, but the UN Conference on the Human Environment in Stockholm in May 1972 met this objection both by embracing the concerns of third world states, and by focusing on the potential limits to global economic growth. This suggested that such growth alone could not meet the needs of the majority of humankind in the third world. Environmental security thus had to be linked to development aspirations, and prospects for international development were necessarily linked to the asymmetric environmental impact of industrialised states.

In the light of these developments, a new consensus had emerged among most peace researchers in the early 1970s, so that by 1973 the editors of the *Journal of Conflict Resolution* sought to broaden the Journal's remit beyond its previous concentration on interstate conflict and the nuclear issue:

---

[15]P. Lawler, *A Question of Values: Johan Galtung's Peace Research* (Boulder CO, Lynne Rienner, 1995), p. 237. Lawler, not himself a minimalist, offers an epistemological critique.

[16]K. Boulding, 'Twelve friendly quarrels with Johan Galtung', *Journal of Peace Research*, 14, 1 (1977), 75–86; K. Boulding, 'Future directions in conflict and peace studies', *Journal of Conflict Resolution*, 22, 2 (1978), 342–54.

[17]Nassau Adams, *Worlds Apart* (London, Zed, 1993).

> The threat of nuclear holocaust remains with us and may well continue to do so for centuries, but other problems are competing with deterrence and disarmament studies for our attention. The journal must also attend to international conflict over justice, equality, and human dignity; problems of conflict resolution for ecological balance and control are within our proper scope and especially suited for interdisciplinary attention.[18]

Another area of controversy within peace research which persists to this day, concerns the notions of violence, conflict and force. This controversy reached its peak at the end of the 1960s and was coincident both with social unrest in Western Europe and with the deep intellectual splits throughout the USA occasioned by the Vietnam War. In a sense, Vietnam crystallized the problem – traditional peace researchers sought a negotiated end to the conflict, but some radicals sought a defeat for the USA and its South Vietnamese client state by the Viet Cong as the only means of achieving justice. Extrapolating from Vietnam, and conscious of the bitter experiences of exploitation in Southern Africa and Latin America, younger and more radical peace researchers argued that the winning of wars might, in some circumstances, be a necessary prerequisite for lasting peace. This echoed some aspects of contemporary 'liberation theology'. They argued, furthermore, that more traditional ideas of seeking negotiated settlements through conciliation and mediation could even be counter-productive by perpetuating structural inequalities in the guise of 'peace'. Such a conflict of views cut deep within peace research, not least because so many of the early peace research initiatives had come from a religious-pacifist tradition, often originating in the 'peace churches' such as the Quakers. In broad terms, most peace researchers, while seeing violence in its various guises as the antithesis of peace, agreed that this did not apply to the concept of conflict. The stated purpose of conflict resolution, for example, was not to prevent conflict – an impossible and also often an undesirable aim – but to transform actually or potentially violent conflict into peaceful (non-violent) processes of political and social change.[19] Indeed, this might increase levels of overt conflict in the short term as marginalized and repressed groups challenged the status quo in the transition to what would eventually become more sustainable and just relations. Where there was, and remains, most dispute was on the question of the use of coercion and force. Here some peace researchers were prepared to acknowledge the legitimacy of military action under certain restrictive conditions, whereas others were not.[20]

By the early 1970s peace research, drawing from a wide range of disciplines and methodologies and with a reasonably sound albeit still restricted institutional base, had defined its subject area in relation to the three great interconnected projects of

---

[18]*Journal of Conflict Resolution,* 27, 1 (1983), p. 5.

[19]Whereas some social scientists, such as those of the Chicago school, regarded conflict as dysfunctional, most peace researchers followed Lewis Coser, in the tradition of George Simmel, in regarding it as functional and intrinsic to human social relationships in general, *The Functions of Social Conflict* (New York, Free, 1956).

[20]For some of the surprising interrelations here, see R. Miller, *Interpretations of Conflict: Ethics, Pacifism and the Just War Tradition* (Chicago, University of Chicago Press, 1991).

avoiding nuclear war, removing glaring inequalities and injustices in the global system, and achieving ecological balance and control. As the institutional base for peace research and conflict resolution continued to expand over the next two decades,[21] so did the main areas of academic interest. We can briefly mention three of them.

First, there was continuing concern with the nuclear threat, but also with chemical and biological weapons and with the dangers inherent in deterrent postures allied to mutual worst-case defence strategies. Progress in arms control, including the Strategic Arms Limitation Talks, were seen to vindicate Osgood's 'graduated reciprocation in tension-reduction' (GRIT) approach,[22] and to exemplify Axelrod's analysis of the 'evolution of co-operation'.[23] Much of this was subsequently thrown into jeopardy by the rapid deterioration in Cold War tensions towards the end of the 1970s and sudden intensification in the nuclear arms race. Although all the existing trends in peace research continued, many centres found it necessary to re-focus resources on the nuclear issue and East–West relations. In effect, they provided an intellectual resource for the burgeoning anti-nuclear movements and rapidly incurred bitter controversy as they were seen by realists and neo-realists, especially within the strategic studies community, as engaging in heavily biased research. This controversy expanded to embrace wider political issues and, in many countries, peace studies and peace research came under persistent attack from centrist and right-wing politicians who labelled it appeasement studies.[24] The vigour of the attacks was at its greatest in the early to mid-1980s and had two unexpected effects: to encourage many able and committed students and researchers to take up peace and disarmament issues, and to improve the quality of the work undertaken as it came under more intense scrutiny. This also made research funding from conventional sources hard to come by, raising difficult ethical questions about the relationship between disinterested scholarship and committed political causes. The Alternative Defence Commission in the UK, for example, an independent body albeit with a university base which analysed non-nuclear defence alternatives over a wide spectrum from techno-fixes to civilian resistance from 1981 to 1987, was dependent almost entirely on Quaker trusts for support, yet within a few years its work was regarded as one of the main sources of critical perspectives on nuclear defence issues.[25]

---

[21]See the institutional time-charts in L. Kriesberg, 'The development of the conflict resolution field', in W. Zartman and J. Rasmussen (eds), *Peacemaking in International Conflict: Methods and Techniques* (Washington, Untied States Institute of Peace Press, 1997), pp. 51–77; and Miall et al., *Contemporary Conflict Resolution,* ch. 2.

[22]C. Osgood, *An alternative War or Surrender* (Urbana IL, Urbana University Press, 1962).

[23]R. Axelrod, *The Evolution of Cooperation* (New York, Basic, 1984).

[24]Caroline Cox and Roger Scruton, *Peace Studies: a Critical Survey* (London, European Institute for Defence and Security Studies, 1984).

[25]The Alternative Defence Commission, *Defence without the Bomb* (London, Taylor and Francis, 1983).

There were, though, some examples of peace research activities receiving support from sympathetic governments. Here, though, there was another problem, as changes in political persuasion could put such initiatives at risk, a difficulty experienced in Germany and Australia in recent years.

A second growth area was in the empirical study of negotiation and mediation processes. The Harvard Program on Negotiation took on the win-win problem-solving and mutual gain vocabulary of conflict resolution, popularized through Fisher and Ury's 1981 *Getting to Yes.*[26] It involved a consortium of academic centres, and, in authentic peace research and conflict resolution vein, drew from a range of disciplines including politics, psychology, anthropology, sociology, and international relations, as well as labour relations, community relations and public planning. A number of other systematic studies were also produced.[27] The same is true in the field of mediation, where Pruitt's complaint about the deficit in critical mediation studies was soon made up.[28] This included attempts to assess the relative efficacy of so-called Track I and Track II mediation approaches.[29] To this should be added continuing efforts to develop the theory and practice of 'problem-solving workshops' in resolving apparently intractable conflicts, notably through the work of Kelman.[30] Reference should also be made to the plethora of initiatives in domestic conflict resolution in family conciliation, labour and community mediation, and Alternative Dispute resolution.

Much of this may be said to have been brought together through a third area of research: the pioneering analytic work produced from the late 1970s on what were variously termed 'deep-rooted conflicts',[31] 'intractable conflicts',[32] or 'protracted social conflicts'.[33] Although barely noticed in the mainstream international relations and strategic studies literature of the time, this can be seen to have anticipated much of the post-Cold War shift of focus to what Rice has called 'wars of the third

[26]R. Fisher and W. Ury, *Getting to Yes* (Boston, Houghton Mifflin, 1981).

[27]In the late 1970s and early 1980s this included, among others, work by Rubin and Brown, Druckman, Zartman and Raiffa.

[28]In the late 1980s and early 1990s this included, among others, work by Zartman, Touval and Zartman, Mitchell and Webb, Kressell and Pruitt, Bercovitch and Rubin, and Princen.

[29]See M. Berman and J. Johnson (eds), *Unofficial Diplomats* (New York, Columbia University Press, 1977); J. MacDonald and D. Bendahmane (eds), *Conflict Resolution: Two Track Diplomacy* (Washington, Center for the Study of Foreign Affairs, 1987); and *Journal of Peace Research,* 28, 1 (1991), Special Issue.

[30]H. Kelman, 'The problem-solving workshop in conflict resolution', in R. Merritt (ed.), *Communication in International Politics* (Urbana IL, University of Illinois Press, 1972); C. Mitchell and M. Banks, *Handbook of Conflict Resolution: The Analytical Problem-Solving Approach* (London, Pinter, 1996).

[31]J. Burton, *Resolving Deep-rooted Conflict: a Handbook* (Lanham, University Press of America, 1987).

[32]L. Kriesberg, T. Northrup and S. Thorson (eds), *Intractable Conflicts and Their Transformation* (Syracuse, Syracuse University Press, 1989).

[33]E. Azar, *The Management of Protracted Social Conflict: Theory and Cases* (Aldershot, Dartmouth, 1990).

kind'.[34] Traditional distinctions between domestic and international politics were rejected as 'artificial' in these cases, and the lack of attention paid to 'ethnic and other forms of communal conflict' by international relations and strategic studies scholars was criticized. Instead, the roots of prevailing patterns of conflicts were traced to the 'disarticulation' between state and society in the post-colonial world, to grievances resulting from deprivation of collective needs, to failures of governance, and to patterns of 'international linkage'. Whether or not this would result in violent conflict in particular cases was seen to depend upon the more contingent actions and events of 'process dynamics' made up of 'communal actions and strategies', 'state actions and strategies' and 'built-in mechanisms of conflict' feeding benign or malign spirals of cooperation or mutual hostility.[35] Remedies were consequently sought in contextual change at international level (for example, via more equitable and accountable global and regional arrangements), structural change at state level (for example, via appropriate constitutional adaptations and the promotion of good governance), relational change at conflict party level (for example, via community relations and reconciliation work) and cultural change at all levels (for example, via the transformation of discourses and institutions that sustain and reproduce violence).

By the end of the 1980s the peace research community had become a diverse but active international group of some two to three thousand scholars and practitioners worldwide, divided by differences of emphasis and approach, but united in the aim of studying how best to overcome threats of large-scale violence and to promote structural change and equitable relations between the myriad collectivities that make up the global community at a time when mounting problems seemed likely otherwise to threaten mutual disaster.

15.3 Peace Research – The Future

How relevant is the peace research agenda today? To answer this question we need to assess briefly what the main future security challenges are likely to be, and then look again at the current state of the peace research field to see if it is equipped to make a significant contribution.

Global security during the 45 years of the Cold War was dominated by the East-West confrontation, but was also a period of major conflicts in many parts of the world with over 100 wars leading to more than 20 million deaths and well over 50 million serious injuries. The ending of the Cold War, while leading to the settlement of some long-standing conflicts, has increased instability elsewhere – the

---

[34]E. Rice, *Wars of the Third Kind: Conflict in Underdeveloped Countries* (Berkeley CA, University of California Press, 1988).

[35]Azar, *Protracted Social Conflict,* ch. 1. For an assessment of Azar's work, see Miall et al., *Contemporary Conflict Resolution,* ch. 3.

enduring disorder in former-Yugoslavia being the most serious European conflict in some decades.

In the immediate future of the next five years, serious conflicts are likely to persist in Central and South-West Asia, South-East Europe and several regions of Africa. The impact of economic turmoil in Asian markets is still difficult to assess, but could lead to mounting political conflict in some countries. The global community, largely through the UN, is likely to react in a faltering and sporadic manner.

Over the next thirty years, two developing parameters are likely to influence global security trends. The first is the deep and enduring inequalities in the global distribution of wealth and economic power, likely to ensure that, within thirty years, one seventh of the world's population controls three-quarters of the wealth, largely but not entirely on a geographical basis. While there have been immense efforts at development, the global picture is one of enduring dis-empowerment and increasing socio-economic polarization.

Furthermore, environmental constraints are likely to exacerbate the effects of human activity on the global ecosystem, making it increasingly difficult for human well-being to be improved by conventional economic growth. The combination of wealth-poverty disparities and limits to growth is likely to lead to a crisis of unsatisfied expectations within an increasingly informed global majority of the disempowered.

Three broad conflict trends are probable. The first arises from a greater likelihood of increased human migration through economic, social and environmental motives. Focusing on regions of relative wealth, this is already leading to shifts in the political spectrum in recipient regions, including increased nationalist tendencies and cultural conflict, especially in western Europe and North America. Such tendencies are often most pronounced in the most vulnerable and disempowered populations within the recipient regions.

Secondly, it is probable that environmental conflict will escalate. This may be local or regional, on issues such as food, land, or water, and global on issues such as energy and mineral resources and transnational pollution, the 1991 Gulf War being an early example.

Finally, and probably most important, competitive and violent responses of the disempowered should be expected within and between states. The recent Zapatista revolt in Southern Mexico, earlier Shining Path action in Peru, and disempowerment responses in North Africa, the Middle East and South Asia are all early examples of a developing trend, not infrequently exacerbated by political, religious and nationalist fundamentalisms. This is linked to underlying historically-conditioned weaknesses in many post-colonial states, struggling as they are to accommodate twin pressures of globalization and fragmentation, and a prey to sectarian and factional exploitation. Increased internal political tensions, particularly secessionist movements, in populous states such as China, India and Indonesia, would have very wide repercussions.

At the beginning of this article we listed seven features that have characterized peace research since its inception in the 1950s. We conclude by reconsidering these in relation to the challenges described above. In the process, we take note of some current critiques of peace research, and suggest that it is how the field responds to these critiques that is likely to determine the quality of its contribution in future.

(a) As a problem-oriented field, peace research is defined in the first place by its core concerns. Although to a friendly critic like McSweeney the 'sudden end of the Cold War has left the formal organization of peace studies bereft of a clear focus',[36] in our view its central *problematique*, as defined for example in the 1973 issue of *Journal of Conflict Resolution* quoted earlier, remains of key relevance 25 years later. Nevertheless, the question of focus is an important one, and, in responding to the major challenges outlined above, peace research will do well to concentrate on three areas where its contribution is already distinctive and significant.
The first is in immediate responses, which include the principal *Agenda for Peace* peace support elements, themselves in part drawn from peace research terminology, such as crisis prevention, peace-keeping, peace-making and the shorter-term elements in post-settlement peace-building.[37] This also involves ethically-acceptable intervention, and regional and global arms control and demilitarization. They are required of an international community of states which shows little evidence of wisdom or leadership and consequently places most responsibility on an under-resourced UN system. While improvements in efficiency and capability must come from within the UN, the NGO role is substantial, especially in the more powerful member states of the UN.
The second area is the longer-term processes involved in conflict resolution and conflict transformation. This is a wide agenda, but now quite well understood and reasonably clearly focused, involving the contextual, structural, relational and cultural elements mentioned earlier with reference to analyses of protracted social conflict.[38] Peck describes the building-blocks of sustainable peace and security as 'well-functioning local, state, regional and international systems of governance, which are responsive to human needs'.[39]
The third area concerns fundamental responses at global level. If the analysis offered above of a polarized, constrained and increasingly unstable world is

[36]B. McSweeney, 'The Ethical Foundations of Peace Research: A Sociological Analysis', paper given at the International Studies Association Annual Convention, Minneapolis, 17–21 March 1998 (unpublished), p. 10.

[37]These terms appear, for example, in J. Galtung, 'Three Approaches to Peace: Peacekeeping, Peacemaking and Peacebuilding' in *Peace, War and Defence – Essays in Peace Research Vol. 2* (Copenhagen, Christian Ejlers, 1975), pp. 282–304.

[38]For a survey of the field today, see Miall et al., *Contemporary Conflict Resolution.* Also, R. Fisher, *Interactive Conflict Resolution* (Syracuse, Syracuse University Press, 1997).

[39]C. Peck, *Sustainable Peace: the Role of the UN and Regional Organizations in Preventing Conflict* (Lanham, Rowman and Littlefield, 1998), for the Carnegie Commission on Preventing Deadly Conflict, p. 45. See also, C. Alger and M. Stohl (eds), *A Just Peace through Transformation: Cultural, Economic and Political Foundations for Change* (Boulder CO, Westview, 1988); L. Kriesberg, *Constructive Conflicts: from Escalation to Resolution* (New Haven CT, Rowman and Littlefield, 1998).

correct, then the issue of rich-poor confrontation is likely to acquire a far greater saliency in future. This will demand a comprehensive rethinking of concepts of security, incorporating unprecedented co-operation for sustainable international economic development and environmental management. This needs to be paralleled by progressive demilitarization linked to the establishment and enhancement of regional and global conflict prevention processes mentioned above. For peace researchers there is now an even greater imperative for them to deepen their understanding of the interconnected problems of international economic relations, the possibilities of sustainable development, and their relationship to security.[40]

(b) The second characteristic feature of peace research is its interdisciplinary nature. This has long been a strength, but also opens peace researchers to the charge of eclecticism and the absence of a distinctive peace research methodology and theoretical base. They have been accused of being insufficiently aware of historical perspectives, wedded to simplistic campaigning which ignores difficult choices, given to woolly concepts such as positive peace or structural violence which could be applied to almost all structures of society, and all too ready to see arms races and militarization as sufficient causes of war.[41] Just though these charges may have been in individual instances, the field as a whole is more substantial than that. We have seen how at least three broad approaches have been combined: a practical problem-solving and needs-based approach; a rational quantitative and comparative-empirical approach[42]; and a theoretical-structuralist approach. It is unrealistic to expect a single peace research methodology or grand theory. What can be expected, however, is continuing high standards of research methodology in these areas, and much greater efforts to increase mutual dialogue with related fields, such as international relations and development studies. It is notable how sparse such dialogue has been in the past, and peace researchers must share a good measure of blame for this.

(c) A third feature is the focus on non-violent processes of political and social change. Here mainstream peace research has been criticized, on the one hand by realists for whom power and coercion is the only international currency, and on the other by neo-Marxist and radical thinkers for whom it is misconceived to attempt to reconcile interests that should not be reconciled. These are large

[40]See, for example, Ben Jackson, 'Promoting real security – implications for policy in the North' in G. Tansey, K. Tansey and P. Rogers (eds), *A World Divided – Militarism and Development after the Cold War* (New York, St. Martin's, 1994), pp. 83–106.

[41]A. Roberts, 'New Peace Research, Old International Relations', in J. Nobel (ed.), *The Coming of Age of Peace Research* (Groningen, Styx, 1991), pp. 1–23.

[42]For example, efforts to quantify correlates of war and large-scale violence since the time of Sorokin, Richardson and Wright; game-theoretic analyses of conflict; and, latterly, empirical studies of negotiation and mediation (see footnotes 26 and 27) and quantitative analyses of the 'democratic peace hypothesis' (see N. Gleditsch and T. Risse-Kappen (eds), Democracy and Peace Special Issue, *European Journal of International Relations,* 1, 4 (December 1995).

issues and subject to continuing debate within the peace research field.[43] In our view peace research must keep its main focus on 'the peaceful settlement of disputes', but also has a significant contribution to make on ways in which military forces, for example in peace support operations, can best play their part within overall peace processes. The neo-Marxist critique is misconceived to the extent that it ignores the strong tradition within peace research which espouses vigorous struggle for social justice, albeit by non-violent means.

(d) What has from the beginning been a refusal to accept the 'internal'/'external' conceptual straitjacket and a focus on multi-level analysis of complex conflict formations in peace research is now less controversial, having latterly been more widely accepted in other academic circles. Perhaps the most significant recent development here has been criticism of earlier concentration on 'top-down' peace-building processes guided by outside 'experts', and insistence on the importance of 'peace-building from below' with a main emphasis on supporting transformative indigenous capacity.[44] This remains an important agenda for the future in which peace research is well-suited to take a leading role.

(e) The original global and cross-cultural aspirations of peace research remain central, but a vigorous gendered critique of past peace research,[45] and the current saliency of the 'culture question' show that these aspirations are in many ways as yet unfulfilled. In addition to its roots in Eastern as well as Western religious traditions, the question of multi-cultural validity became important in peace research in the 1960s through the influx of anthropological perspectives,[46] and erupted into a major controversy in the 1980s in the form of critiques of universalist or 'generic' (and, in particular, 'western') conflict resolution models.[47] Recent comparative studies of culturally diverse ethno-conflict theory and ethnopraxis in African, Latin American, Asian and Islamic countries points the way ahead.[48] In our view this is perhaps the most important single challenge in peace research today – its evolution into a fully global and cross-cultural venture.

---

[43]P. Wehr, H. Burgess and G. Burgess, *Justice without Violence* (Boulder CO, Lynn Rienner, 1994).

[44]J. Lederach, *Building Peace: Sustainable Reconciliation in Divided Communities* (Washington DC, United States Institute of Peace, 1997).

[45]A. Harris and Y. King (eds), *Rocking the Ship of State: toward a Feminist Peace Politics* (Boulder CO, Westview, 1989); A. Taylor and J. Miller (eds), *Conflict and Gender* (Cresskill, Hampton, 1994).

[46]R. LeVine, 'Anthropology and the study of conflict: an introduction', *Journal of Conflict Resolution,* 5, 1 (1961), 3–15; M. Ross, *The Culture of Conflict: Interpretations and Interests in Comparative Perspective* (New Haven CT, Yale University Press, 1993).

[47]See, for example, K. Avruch, P. Black and J. Scimecca, *Conflict Resolution: Cross Cultural Perspectives* (Westport, Greenwood, 1991).

[48]For example, P. Salem (ed.), *Conflict Resolution in the Arab World: Selected Essays* (New York, American University of Beirut, 1997).

(f) The sixth distinctive feature of peace research, its combination of objective analysis and normative commitment, can also be seen as both a strength and a difficulty. The tension between aspirations to build a 'science' of peace in which the analysis of violent conflict could be likened to the analysis of disease in the medical sciences, and the ethical political goals inherent in non-positivistic moral choices at the heart of the peace research project, is a creative one, but has also been seen by some as academically suspect. Here peace researchers would do well to accept the full implications of the ethical basis of the subject, along the lines suggested by McSweeney,[49] acknowledging constructivist critiques, and establishing clearly that, while peace research does not prescribe specific solutions or end goals for society, its commitment to the search for non-violent processes of political change implies deep transformations in existing power structures.

(g) Finally, there is the close relationship in peace research between theory and practice, once again controversial academically, but a strong and distinctive tradition. In addition to the problematic but fruitful relationship between peace research and peace activism, there is the characteristic way in which, when all goes well, practice informs theory and theory is properly questioned as to its practical implications. This applies, for example, to current attempts to apply critical theory within the field.[50] With reference to the three traditional approaches outlined above, neither the 'subjectivist' problem-solving approach, nor the 'objectivist' rationalist-empirical approach are considered adequate to the core issues of ethical choice and social action, since both assume purposive agents and utility-maximising decision-makers and ignore constitutive constraints. The theoretical 'structuralist' approach, on the other hand, is seen to ignore constructivist insights and to fail to account for the way social contradiction transmutes or does not transmute into violent conflict. From this perspective the danger of failing to adopt a critical-theoretic approach which bridges the ontological gap between agent and structure and adequately questions the institutional, discursive and practical origins of exclusion and violence, is that otherwise well-intentioned interventions are likely simply to reproduce them. From a peace research perspective one crucial value test of this approach will be its pragmatic pay-off in terms of better insights into practical problems and improved responses. Here again there are hopeful signs.

[49] McSweeney, 'The ethical foundations of peace research'.

[50] See, for example, V. Jabri, *Discourses on Violence: Conflict Analysis Reconsidered* (Manchester, Manchester University Press, 1996).

15.4 Conclusion

Faced with the challenges outlined above, and equipped with what is now a 50-year tradition of applied analysis, peace researchers have a more complex and difficult role than the relatively straightforward activities of the Cold War era, but one that is essential if scholarship is to contribute with any effect to achieving a more genuinely just and peaceful world. Hampson has identified 'four different schools of intervention' in the international community's responses to contemporary violent conflict: hard realism and soft realism which have a 'state-security orientation', psychological approaches which have a societal or human security-based orientation, and governance-based approaches which focus on state-society linkages.[51] In its historical concerns, peace research has things to say about the latter three. Our overall conclusion, therefore, is that, so long as the response to current critiques is positive and creative, inherited traditions have equipped the peace research community with a corpus of knowledge, outlooks and attitudes which is well suited to responding to the developing problems of insecurity and conflict which are likely to be such a dominant feature of the early twenty-first century.

[51]F. Hampson, 'Third-party roles in the termination of intercommunal conflict', *Millennium,* 26, 3 (1997), 727–50.

Chapter 16
A Century on the Edge (2014)

Contributions to anniversaries may take a broad view, sometimes even a centuries-long perspective.[1] Most commonly, such perspectives are framed within traditional boundaries – nineteenth, twentieth or twenty-first century, for example – but this article argues that one period not normally viewed in this way, from 1945 to 2045, may have a particular significance, giving a perspective that is of value in analysing current trends in international security.

The argument is that in a broad view of human evolution within a largely closed planetary system, this century has a peculiar relevance in that it is likely to cover the period when humankind must learn to live with two dangerous capacities. One is the development of military technologies that, if used, could set back the human community many decades, possibly centuries, and the other is the impact of anthropogenic effects on global ecosystem homeostasis.

In broad terms, the article argues that the first element is slowly being understood, but that success in avoiding the intercontinental use of weapons of mass destruction has been more a matter of luck than judgement or wisdom. This element predominates during the first two-thirds of the 1945–2045 period, more or less through to the time of writing. In respect of the second element, responding to global environmental constraints, the prognosis is currently poor. This question dominates the latter two-thirds of the century, obviously overlapping with the first. The article argues that there are significant connections between the two issues, and that learning effectively from the inadequate response to the first issue may assist in formulating a better response to the second.

Before proceeding further it is worth pausing to consider a much longer perspective. Tool-using pre-hominids evolved several million years ago and, as tool-making became a feature of early hominid society, the species spread across the Earth. Even 15,000 years ago, though, the worldwide population was little more

[1]This text was first published as: "A century on the edge: from Cold War to hot world, 1945–2045", *International Affairs*, 90:1 (2014) pp. 93–106. The permission to republish this text was granted on 28 February 2018 by Oxford University Press.

P. Rogers, *Paul Rogers: A Pioneer in Critical Security Analysis and Public Engagement*, Pioneers in Arts, Humanities, Science, Engineering, Practice 21,
https://doi.org/10.1007/978-3-319-95150-8_16

than 5 million as hunter-gatherers require substantial areas of territory to provide sufficient ecological carrying capacity to satisfy their needs, principally for food. The agricultural revolution that saw the domestication of plants and animals established in several separate regions across the world revolutionized the ability to provide food, increasing the land's carrying capacity and leading both to an explosion of the human population and to the emergence of towns and cities.

More recently, a second period of revolutionary change commenced with the start of the industrial revolution, accompanied by linked agricultural developments, some 250 years ago, followed by further population growth and urbanization. This process, which continues today, has involved substantial environmental impacts, most commonly at local and regional levels, with little evidence of impacts on the entire biosphere until the late twentieth century. In its wake have come numerous developments in military technology, starting with the mechanized warfare of the early part of the twentieth century and culminating in the nuclear age. The relevant point for present purposes is that, if measured in terms of broad human history, the huge changes of the last 250 years are very sudden, and the acquisition of capabilities for self- and environmental destruction even more recent.

In examining the century from 1945 to 2045, this article starts by summarizing some of the features of the nuclear arms race of the Cold War before examining how military postures changed after the Cold War ended and were later bound up in the response to the 9/11 attacks and subsequent reorientations of that posture. It then analyses the issue of environmental limits on human activity, especially climate disruption, but does this in the context of the socio-economic divisions that have deepened over the past three decades. Finally, it seeks to draw on the experience of the Cold War confrontation in suggesting responses to future challenges that might enhance equality, sustainability and security.

16.1 The Cold War Experience

Following the origins of nuclear weapons in the US Manhattan Project, efforts to curtail their further development in the late 1940s failed and by 1955 the United States and the Soviet Union had between them over 3,000 nuclear weapons, rising to 37,000 in 1965 and 46,000 in 1975, and peaking at over 62,000 around 1985. In the early years of the arms race the rival states placed some emphasis on very powerful weapons with explosive yields of up to 25 megatons, some 2,000 times the power of the Hiroshima bomb. Later, the emphasis was on very large numbers of smaller weapons, although many of these were individually far larger than the Hiroshima bomb.

At the strategic level, there were interlocking arms races involving a triad of weapon types – intercontinental ballistic missiles, long-range bombers and sea-launched ballistic missiles – with each side developing increasingly accurate systems leading to a perceived risk of vulnerability and hence to launch-on-warning and launch-under-attack options. By themselves these raised particular problems of potential crisis instability stemming from the risk of a 'use them or lose them' mentality, but this was made worse by the proliferation of weapons types.

As well as a strategic arms race, with each side maintaining this triad, there were numerous types of theatre and tactical nuclear weapons, with a view prevailing among the nuclear planning fraternity that small nuclear wars could be started and controlled without escalation to a central nuclear exchange. NATO consistently maintained (and indeed still does) a preparedness to use nuclear weapons first codified in 1967–8 in MC14/3, *Overall Strategic Concept for the Defence of the NATO Area*, and while the Soviet Union denied having a policy of first use, few in the West took this seriously.

Nuclear escalation was more easily envisaged as nuclear weapon deployments permeated almost every part of the NATO and Soviet force structures. Chemical weapons did too, ranging from the maintenance of stocks of long-established agents such as mustard gas through to potent nerve agents including Sarin, Tabun and V-agents, with delivery systems including strike aircraft, ballistic missiles and artillery. At the peak of the Cold War the United States had 27,000 tons of chemical agents and the Soviet Union 40,000 tons, but even with these arsenals the potential for destruction was less than nuclear weapons and both states are now slowly eliminating their stocks under the terms of the Chemical Weapons Convention.

For nuclear weapons, though, the penetration of types was quite astonishing and is largely unrecognized by new generations of scholars and analysts. The weapons included free-fall bombs, stand-off missiles, anti-submarine depth bombs, torpedoes, air-to-air missiles, battlefield weapons, artillery, mortars and even small 'backpack' nuclear mines for destroying tunnels and bridges. Some had explosive forces of much less than a kiloton (1,000 tons of high explosive-equivalent) meaning that the threshold between very small tactical nuclear weapons and the largest fuel-air-explosive conventional bombs was remarkably small. This combination of a first use posture with the deployment of nuclear warheads at quite low levels of command structures, together with extensive planning on both sides for limited nuclear war, was an element of the Cold War that contrasted markedly with a public perception of stable deterrence.

Looking back now, nearly a quarter of a century after the Cold War ended with the collapse of the Soviet bloc, there is still a common view that nuclear weapons kept the peace and provided overarching stability; but there are serious problems with this assertion. One is that while a central nuclear exchange did not happen, numerous proxy wars were fought that reflected Cold War competition. During the period from 1950 to 1990, they included conflicts that killed over 70,000 people in both Nicaragua and El Salvador, 390,000 in Angola, over 600,000 in the Horn of Africa and over 1 million in Mozambique. Losses in Asian proxy wars were even greater, with 1.3 million dying in Afghanistan, 2.3 million in Vietnam and 3 million in Korea.[2] Many millions of people were seriously wounded, often with

[2]Ruth Leger Sivard, *World military and social expenditure 1991* (Washington DC: World Priorities, 1991).

life-changing impacts, and the material and economic damage to societies had effects stretching over more than a generation. Overall, the Cold War proxy conflicts killed at least 10 million people – a fact that has to be set against any notion that nuclear weapons kept the peace.

Furthermore, many studies over the last 25 years have questioned the idea of stability achieved through nuclear deterrence. Because of the relatively closed nature of Soviet society, little is known even now of the nuclear accidents and crises that affected its forces, but it is a reasonable assumption that they were no less dangerous than those experienced by the United States. These include many examples of serious accidents involving nuclear weapons, in the course of which some nuclear weapons were lost and never recovered. Certain crises were so severe that they later contributed to some of the key people involved working towards a nuclear-free world, a notable example being President Kennedy's Defense Secretary, Robert McNamara, who was much influenced by post-Cold War discussions with former Soviet counterparts from the period of the Cuban Missile Crisis.[3] This was shown, after the Cold War, to have come far closer to a central nuclear exchange than was realized at the time, not least because of each side's ignorance of the other's conventional and nuclear deployments and intentions. In another remarkable case, NATO's 1982 *Able Archer* exercise involving highly accurate road-mobile nuclear missiles resulted in an immediate escalatory Soviet response as elements within the military leadership feared the operation was a preparation for an actual nuclear attack. The impact on NATO planners was to modify their entire exercise programme, though knowledge of this only came into the public domain many years later.

The proxy wars and risks of nuclear disaster are paralleled by a third aspect of the Cold War period: the extraordinary commitment of human ingenuity, research prowess and financial resources to the East-West confrontation. To take just one example, in the early 1970s there was a serious world food crisis that seemed likely to lead to a period of famine stretching across much of the tropics and sub-tropics and putting the lives of tens of millions of people directly at risk. The worst aspects of the short-term problems were overcome by 1975 through welcome international cooperation, but there were also concerted efforts to promote spending on agricultural research to prevent a repeat of the crisis in the future. This involved calls for an increase in agricultural research directed towards the global South from approximately US$1.5 billion (at 1974 prices) to US$5 billion for a period of ten years. This was equivalent to around 2 per cent of world military spending at the time.[4] The increase in research spending never came near that laudable target and, nearly 40 years later, the number of under-nourished people is actually higher.

---

[3]Robert S. McNamara, 'The conference on disarmament should focus on steps to move towards a nuclear-free world', *Disarmament Diplomacy* 4: 2, April 1996, pp. 8–10.

[4]Paul Rogers, *Food in our time—but not just yet: a report on the World Food Conference* (London: World Development Movement, 1974).

This is one specific example of contrasting attitudes to civil and military research and development but there are many others. Britain is an interesting case in point if one looks back on the early Cold War period of the 1950s. At this time, UK military R&D did not just involve the development of nuclear and thermo-nuclear weapons but the UK actually developed and then deployed three different types of transcontinental bomber, Valiant, Vulcan and Victor, a range of offensive and defensive missiles and numerous other conventional army, navy and air force systems while maintaining armed forces close to four times the current size.

Looked at overall, the Cold War period was a time of great risk in which an intercontinental catastrophe was avoided not by wisdom but with luck, and during which human resources were wasted on an immense scale. Why highly intelligent people – military, politicians, diplomats and others – were unable to face up to the dangers of the nuclear environment in which they were willing to operate is still difficult to understand. Nash's seminal essay, 'The bureaucratization of homicide',[5] is informative as to the outlooks of individuals lower down the chain of command, pointing to behaviour within any bureaucracy. In his experience, diligent nuclear planners saw it as an essential part of their work to be effective, not least as a means of responding to those above them and moving up the career ladder and the higher levels of security clearance. If and when they did consider the impact of what they were doing, they could fall back on the certainty that their Soviet opposite numbers were similarly engaged and that their own endeavour was driven by patriotic duty.

This throws little light on the roles of the most senior actors. Military commanders below the top tier were commonly engaged in maintaining their own forces, and their most senior commanders were hugely influenced by the Cold War ideological divide and the threat from their opponents, a threat that also served the interests of powerful defence lobbies on both sides. As to the most senior politicians, they typically served quite short terms and were engaged in their own political environments that saw any reaching out to opponents as signs of weakness. Occasionally, as with the Kennedy outreach to the Soviets after Cuba, and Gorbachev's openness from 1995, individuals rose above the immediate confrontation. With Kennedy it did not outlast his death and in Gorbachev's case he may have played a huge role in ending the Cold War but that is not sufficiently acknowledged more than two decades later.

That such a dangerous arms race was able to continue for several decades should not disguise the fact that its ending at the close of the 1980s heralded a period of decreased nuclear danger, with some of the most dangerous and potentially unstable 'first use' systems being downgraded as part of a process of substantial overall reductions in arsenals. Nearly 25 years later, there is a much decreased risk of an intercontinental nuclear war but a slowly increasing problem of proliferation. The five acknowledged nuclear powers who occupy permanent seats on the UN Security Council, the United States, Russia, China, the UK and France, are all maintaining

---

[5]Henry T. Nash, 'The bureaucratization of homicide', *Bulletin of the Atomic Scientists* 36: 4, April 1980, pp. 26–34.

and even developing their nuclear systems albeit, with the exception of China, mostly at much-reduced levels; outside this group, Israel has been joined by India and Pakistan as nuclear powers, North Korea has a limited nuclear capability, and there remain risks of proliferation in the Middle East depending on the outcome of diplomatic interchanges between Iran and the United States.

Despite deep reductions in nuclear arsenals in the US, Russia, France and the UK, progress towards nuclear elimination has been very limited and this does not give confidence that leaderships have the ability to avoid future technological dangers, whether these stem from bio- or nanotechnologies, or other developments. Genetic manipulation, for example, has the potential to add substantially to the effectiveness of biological agents; yet, while there is a worldwide ban on such developments, attempts over two decades to strengthen the 1972 Biological and Toxin Weapons Convention have led to little in the way of inspection or verification procedures. Similarly, the pace of development in nanotechnologies has not been accompanied by sufficient consideration of untoward impacts.

On the evidence to date, the first two-thirds of the 1945–2045 century does not inspire confidence that the human community has acquired sufficient wisdom to handle the destructive potential of some key scientific and technological developments, even if catastrophe has so far been avoided. Furthermore, the twelve years since the 9/11 attacks suggest that vigorous recourse to military force is the most likely response to new threats, what might be termed the 'control paradigm', a concept most clearly demonstrated by analysing the response to those attacks.

16.2 A Jungle to Be Tamed

When President Clinton's newly appointed Director of Central Intelligence, James Woolsey, was questioned in confirmation hearings on Capitol Hill in 1993, he was asked to characterize the transition to the post-Cold War world: he replied that the dragon had been slain but the United States now inhabited a jungle full of poisonous snakes. In the following eight years US armed forces re-orientated their posture away from the Cold War and towards taming that jungle. As well as scaling down the nuclear capability, there was a move away from large-scale armoured forces and anti-submarine capabilities, whereas expeditionary warfare, long-range strike capacity and stand-off weapons were largely maintained. Even with nuclear weapons the drawdown in number did not mean a move away from nuclear postures in general, more their adaptation to a potentially proliferating world rather than a major East-West confrontation. Moreover, the concern of Putin's Russia with rebuilding and enhancing its nuclear forces would appear to have much to do with seeking to regain superpower status.

With the election of George W. Bush to the presidency in 2000 and the appointment of Donald Rumsfeld as Secretary of Defense the following year, this trend in US military posture continued, even though the overall security orientation was greatly influenced by the neo-conservative vision embodied in the 'Project for a

New American Century'. That vision required singularly strong armed forces; but while Rumsfeld bought heavily into the idea of using military power to maintain control, he saw 'war lite' as being the desirable approach – further developing the trends of the 1990s. On the nuclear front the new administration was particularly concerned with maintaining nuclear capabilities, a key aspect of this being opposition to the 1972 Anti-Ballistic Missile Treaty. More generally, robust and versatile forces able to intervene at short notice wherever US interests were threatened were considered appropriate to a world in which a single super-power would be able to guarantee civilization while running, in Krauthammer's phrase, a 'benign imperium'.[6]

The initial response to the 9/11 atrocities followed this pattern, with termination of the Taliban regime achieved by a combination of air power, special forces, and the use of Northern Alliance militias as surrogate ground troops. Eighteen months later, with the 'war on terror' having been expanded to encompass an 'axis of evil', the termination of the Saddam Hussein regime in Iraq was achieved with less than a third of the forces employed to evict the Iraqis from Kuwait twelve years earlier. By 1 May 2003, the occasion of President Bush's 'mission accomplished' speech from the flight deck of the USS *Abraham Lincoln*, the President could point to the termination of the Taliban regime, the dispersal of the Al-Qaeda movement, the prospect of Afghanistan and Iraq developing as strongly pro-western states and greater US influence in the geostrategic Central Asian region in the wake of new base agreements.

Perhaps most significant of all were two elements relating to Iraq. One was that the state's economy would be remodelled under Paul Bremer's Coalition Provisional Authority to evolve into a true free market – a beacon for the wider Middle East and a model for the New American Century. The second was that the most substantial remaining problem for the United States in the region, the Islamic Republic of Iran, would now be thoroughly constrained. This might be difficult to understand, given that Iran's enemies – Iraq and the Taliban – were also enemies of the United States and that Iran had actually been helpful to the US in its actions in Afghanistan immediately after the 9/11 attacks. Part of the explanation relates to Iran's antagonism to Israel but it also goes back to the Islamic Revolution and the loss of the Shah's Iran as a bulwark for the United States in the wider Middle East. In this context, therefore, the establishment of US bases in Afghanistan and Iraq, pro-American allies in western Gulf States and the US Navy's Fifth Fleet in the Gulf and the Arabian Sea would all ensure a compliant and subdued Tehran. War lite would work.

In the event, the outcome of these military actions was formidably different from expectations, and by 2010 two bitter and prolonged wars had resulted in over 200,000 people killed, hundreds of thousands injured, nearly 8 million displaced

[6]Charles Krauthammer, 'The Bush Doctrine: ABM, Kyoto and the new American unilateralism', *Weekly Standard*, no. 36, 4 June 2001.

and a massive expenditure likely to reach US$3 trillion.[7] As the security situation in Iraq deteriorated, Barack Obama was able to campaign for the presidency in 2008 on the basis of an increasingly unpopular war, and having taken office ended the occupation in Iraq by 2011. The war in Afghanistan was maintained during his first term before reality intervened. What had started as 'war lite', first in Afghanistan and then Iraq, had become an unsustainable 'war heavy', with over 100,000 troops deployed in each country yet in neither place succeeding in achieving their objectives.

Moreover, at the centre of the original 'war on terror' had been a determined effort to destroy the Al-Qaeda movement; and this also failed. Even the dispersed and weakened movement was undergoing a metamorphosis into a multi-faceted entity with numerous allied groups and individuals. Much of the impact was experienced within the Islamic world, not least in Iraq and Pakistan with many thousands of people killed, but over the period 2002–2007 there were also many attacks on western and Israeli targets across the world, in addition to escalating violence in Afghanistan and Iraq. They included attacks in Casablanca, Djerba (Tunisia), Istanbul, Riyadh, Sinai, Aqaba, Amman, Bali, Jakarta, Islamabad, Karachi, Mumbai, Mombasa, Madrid and London, as well as many failed attempts.

By 2010 the United States and its coalition partners were even losing the will to maintain the occupation of Afghanistan, and following his re-election in 2012 President Obama proceeded with the plan to withdraw most troops from the country by the end of 2014. It was now clear that both wars had demonstrably failed to achieve their objectives. Iraq was hardly a part of a brave New American Century—indeed, by late 2013 it was close to civil war—and there were few analysts who believed that the Taliban would have no role in a post-occupation Afghanistan. The 'war on terror' had failed in all its major aims excepting the occurrence, so far, of another serious attack on the continental United States.

16.3 Remote Control

With the abject failure of 'boots on the ground' to bring closure to the two wars, the whole concept of large-scale expeditionary warfare has come into question, partly because of the manifest failure of the wars in Afghanistan to achieve their aims but also because of the increasingly 'casualty-averse' mood in the United States and its coalition partners, especially Britain. This has coincided with developments in military technology and posture which look towards the control of threats through other means. In the popular mind, the main focus of this has been the rapid growth in the use of high-endurance long-range armed drones, deployed increasingly in

---

[7]Eisenhower Research Project, *The costs of war* (Providence, RI: Watson Institute for International Studies, Brown University, 2011); Joseph Stiglitz and Linda Bilmes, *The three trillion dollar war* (London and New York: Allen Lane, 2008).

Afghanistan and credited with inflicting substantial damage on the middle-ranking cohorts of the Al-Qaeda movement centred on tribal areas of north-west Pakistan.

The boots on the ground are thus being replaced by 'remote control'; but this is only partly a matter of armed drones. Other components are the much greater use of low-profile special forces operations, most notably the 'night raids' employed repeatedly in Afghanistan. Such is the expansion of special forces within the US military system that US Special Operations Command has expanded from 42,743 personnel in 2008 to 63,650 in 2012 and is planned to peak at 71,000 by 2015.[8] To give a sense of perspective, this will not be far short of the size of the entire British army after the current programme of cuts to the latter is implemented.

There are three other elements of remote control. One is the growth in private military and security companies, with as many as 20,000 personnel contracted into Iraq and Afghanistan by this route,[9] and a second is the rendering or kidnapping of selected high-value suspects for interrogation, often to third countries, with or without the sanction of the countries from which they are taken. Finally, there is the development over the longer term of Prompt Global Strike—the ability to deliver conventional warheads over intercontinental distances in much less than an hour. Early variants have been based on very large ballistic missiles originally intended for nuclear delivery but retrofitted with conventional warheads. These are likely to be superseded by specifically developed boost-glide systems and hypersonic vehicles using ramjet and scramjet propulsion.[10]

16.4 Lessons Unlearnt

While this trend towards remote control has been under way for much of the past ten years, it appears to have had little more effect on the perceived security challenges than 'war heavy'. In October 2013 the new head of the UK Security Service (MI5), Andrew Parker, could point to a security risk stemming from 'thousands' of radical Islamists in the UK, and there was a clear sense that Al-Qaeda had further evolved into a phenomenon that was more akin to an idea than an effective paramilitary movement. Syria was now the main focus of extreme Islamist paramilitaries, followed closely by Iraq. The tribal areas of Pakistan might now be far behind in potential, but Yemen remained of concern, as did Somalia and parts of the Maghreb, the former in relation specifically to the Westgate Shopping Mall attack in Nairobi the previous month. Of particular concern were western areas of the Sahel region, most notably the actions of the highly active Boko Haram

---

[8]Linda Robinson, *The future of US special operations forces*, Council Special Report no. 66 (New York: US Council on Foreign Relations, April 2013).

[9]'Bullets for hire', *The Economist*, 17 Nov. 2012.

[10]'Speed is the new stealth', *The Economist*, 1 June 2013; Caitlin Harrington, Sam LaGrone and Daniel Wasserbly, 'Silver bullets: US seeks conventional weapons with a global reach', *Jane's International Defence Review* 43: 9, Sept. 2010, pp. 50–53.

insurgency of northern Nigeria, while the Russian authorities were committing huge resources to safeguard the 2014 Winter Olympics in Sochi from attack by the Caucasus Emirate.

While the move towards remote control of threats is initially attractive, it is already being seen to have unexpected consequences. Armed drones are being actively developed by many countries including China and Iran, and precedents have been set by the United States that will make it difficult if not impossible to counter cross-border incursions in pursuit of national security concerns. If the United States can target people for assassination in Yemen, Pakistan or Somalia, why could not Russia do the same in Central Asia, China in South-East Asia or Turkey against Kurdish paramilitaries in Iraq if any of these states believed its security was threatened?

Taking a broad view, the experience of the twelve years from late 2001 to the end of 2013 was hardly satisfactory from a western military perspective, with neither military occupations nor remote control having the desired outcomes. It is in this context that it is now appropriate to analyse the most significant security challenges of the next 30 years – the final third of the century under review – to see whether the experience of those twelve years, and the longer-term experience of the Cold War, bodes ill or well for greater international peace and stability.

16.5 New Challenges

While the industrial revolution resulted in major environmental impacts, these were local and regional in scale rather than affecting the biosphere as a whole. In the early 1970s global systems analyses were undertaken that suggested that world-wide ecosystem homeostasis would at some stage be affected by human activity, the most cogent analysis, *Limits to growth*, being published in 1972.[11] That study argued not that damaging change was inevitable in the short term but that within 60 years profound changes would be under way, the fundamental risk being that the ecological carrying capacity of the whole-planet ecosystem would be substantially exceeded.

The argument was not accepted by mainstream economists and others, and there were also expectations that technological developments would substantially increase carrying capacity. Furthermore, the rapid oil price rises of 1973–4, followed by a period of stagflation, resulted not in a reconceptualization of economic systems along sustainable lines but in the onset of a transition to a far stronger free market outlook which had a substantial impact across the world, not least through the evolution of the Washington Consensus.

A decade later, a 'marker' for global human effects was discovered in the form of CFC pollutants affecting the ozone layer in the upper atmosphere, the first clear

[11]Donella H. Meadows, Dennis L. Meadows, Jorgen Randers and William W. Behrens III, *Limits to growth* (London: Earth Island, 1972).

evidence of a specific human activity having a truly global impact. The critical nature of the threat of ozone layer disruption, combined with the relative ease of moving to alternatives, ensured that intergovernmental action was taken rapidly, resulting in the Montreal Convention of 1987.

Meanwhile, concerns were raised about a far more serious and deeply embedded element of global environmental impact—the potential for increases in concentrations of atmospheric carbon dioxide and methane to cause climate disruption. Although there is an overwhelming scientific consensus on this issue, little governmental action has been taken, and 'climate denial' has been deeply embedded in political and economic circles, most notably in the George W. Bush administration in the United States from 2001 to 2009. On current trends, action taken to curb carbon emissions will be utterly inadequate to counter climate disruption; and this failure to take adequate steps is occurring against a background of deeply asymmetric impacts. Apart from an accelerating impact in near-Arctic regions, climate disruption is expected to be particularly intense across the tropics and sub-tropics, with above-average increases in temperature and changes in rainfall patterns leading to a marked decrease in the ecological carrying capacity of croplands that provide food for the majority of the world's people.

In parallel with the development of this challenge, the global socio-economic divide is widening. Following the deregulation of financial systems in the late 1980s, commencing with the 'Big Bang' in London in 1986, world economic growth continued, albeit at a slower rate than before, but the experience of the past 30 years is that there has been a disproportionate concentration of wealth in barely a fifth of the global population, a large majority of the world's wealth and annual income—close to 85 per cent—being shared by 1.5 billion people out of a world population of 7 billion.[12] This constitutes a transglobal elite, including hundreds of millions of people in China, India, Brazil and other developing and transitional economies, even if largely still concentrated in the countries of the North Atlantic community.

Alongside this substantial elite exists a marginalized majority that is failing similarly to benefit from the current world economic system. According to the Asia Development Bank's *2012 Asian Development Report*, for example, if there had been more even distribution of the fruits of growth, 'another 240 million people in the 45 countries that make up developing Asia would have moved out of poverty in the last two decades',[13] and the World Bank reports that 400 million children under 13 years of age are living in extreme poverty.[14]

---

[12]James Davis, Susanna Sandstrom, Anthony Shorrocks and Edward N. Wolff, *The world distribution of household wealth*, WIDER Angle no. 2 (Helsinki: World Institute for Development Economics Research, 2006).

[13]Neena Bhandari, 'Rising inequality could be Asia's undoing', Terra Viva-Inter Press Services, New York, 13 April 2012.

[14]Jim Lobe, '400 million children mired in extreme poverty', Terra Viva-Inter Press Services, New York, 13 Oct. 2013.

Yet because of the size of the successful elite, it acts largely as a self-contained but global entity which benefits from material well-being that is largely taken for granted. The elite minority world exists alongside the majority world but is hardly conscious of the very existence of the divide. Rarely does it acknowledge the sustained benefits it gains from the appallingly low wages and abysmal working conditions of hundreds of millions of people producing cheap goods.

At the same time, there have been impressive and thoroughly welcome improvements in education, literacy and communications in the past 40 years across the global South, serving to make many members of that marginalized majority all too aware of their own marginalization.[15] The bitterness that this creates is reflected in much of the motivation for the 'Arab Awakening' over the past three years, and in the growth of radical movements such as the Naxalite rebellion in India and social unrest in China. It frequently adds further motivation to extreme groups rooted in religious or nationalist ideologies, the Boko Haram movement's recruiting from marginalized youth in north-east Nigeria being one of many examples.

16.6 Consequences and Responses

There are analysts and policy-makers who are concerned at the persistence of these deep inequalities in human development, and there are others who separately see the prospect of climate disruption as a fundamental threat to human well-being. What needs to be done is to integrate these phenomena more fully and see them as a single issue – the challenge of a world with embedded and deepening socio-economic divisions that is also facing severe environmental constraints. This is the key security issue of the final third of the 1945–2045 century, and unless it can be resolved there is every risk of the world being deeply unstable and insecure by the end of that century.

As climate disruption diminishes the ecological carrying capacity of many of the world's most important croplands and as severe weather events increase in frequency and intensity, the pressures on social and political systems become more serious, and will be reflected in greatly increased migratory pressures and the rise of radical movements – 'revolts from the margins'. As Edwin Brookes wrote 40 years ago, we have to avoid 'a crowded, glowering planet of massive inequalities of wealth buttressed by stark force yet endlessly threatened by desperate people in the global ghettoes'.[16] This will not be easy to do, given the persistence of the control paradigm, in spite of its evident recent failure in responding to the 9/11 attacks. Part

---

[15]David Sogge, 'Inequalities and organised violence', Norwegian Peacebuilding Resource Centre policy brief, Oslo, Sept. 2013.

[16]Edwin Brookes, 'The implications of ecological limits to development in terms of expectations and aspirations in developed and less developed countries', in Anthony Vann and Paul Rogers, eds, *Human ecology and world development* (London and New York: Plenum, 1974).

of the problem lies with the need to make major policy choices over the coming decade that may well have an impact on the minority elite. These choices include the requirement to move rapidly towards ultra-low-carbon economies while also making economic changes that result in greater equity and emancipation. In the face of what seem to be massive changes, it is far easier in the short term to believe that the status quo can be maintained and resort to 'Liddism', keeping the lid on problems rather than addressing the forces underlying them.

Thus we persist in the belief that the liberal market is wholly functional and that the world economic system as it has evolved, now embracing even China, is without question the only way to operate. Even the 2008 financial crisis is receding safely into the past, and when threats to this successful system arise they are dangerous and must be countered. This may well include vigorous support for elite regimes facing revolts from their margins, including the training and equipping of necessary police and special forces. It involves stringent control of migration, intervention in failed and failing states when they threaten our interests, and sometimes even extends to the violent termination of regimes deemed to threaten the security of the established system – 'our' world.

An appropriate narrative is not difficult to promote. Thus economic migrants are after 'our' jobs and asylum seekers are merely scroungers. When 350 people drowned off the Mediterranean island of Lampedusa in October 2013, initial shock soon gave way to indifference. If people riot in cities in the West, they are criminals pure and simple and there is no need for further discussion. If rebels oppose a friendly government in the South they are terrorists, a dangerous threat to established order to be repressed with all necessary force. It is, in short, a matter of taming Woolsey's 'jungle full of snakes', secure in the belief that it can be tamed and order ensured. Moreover, that taming can be effected, at least in part, by remote control, a necessary substitute for the recently failed endeavours of boots on the ground.

This narrative relates to the discussion earlier in this article of attitudes within military and political communities during the Cold War. The possibility of an utter catastrophe was embedded in the very development of extraordinarily powerful weapons of mass destruction together with the active planning for their use, and there were instances where society came far closer to such disaster than was known in the public domain at the time. Even so, in those circles that were aware of the risks, planning was still undertaken: for the first time in human history there existed the capacity for self-infliction of catastrophe, yet the wisdom to respond to it was lacking and disaster was not easily avoided.

16.7 Choices

Over the period through to 2045 it will be wise for political systems to respond to the emerging issue of an economically divided and environmentally constrained world. In one sense it is more difficult to do this than to respond to the nuclear issue, because the effects are incremental rather than having sudden catastrophic potential. In another sense, though, the very fact that they are incremental means that there is more scope for responding. Indeed, such response may even be aided by the 'canary' element, such as the impact of extreme weather events now acting as a herald of more substantial problems to come.

New thinking and analysis are needed to facilitate the development of a sustainable security approach before it becomes an absolute necessity in the face of what might in due course be an unstable world system. An appropriate definition of prophecy in this context might be 'suggesting the possible': that is, proposing the policies and approaches most likely to aid a transition to a more emancipated and sustainable world system.

If that is to happen, then the period through to 2025 is particularly important, as fundamental changes are already beginning and will be having their impact by 2045. Reference was made earlier to two transitional periods in human history, the agricultural and industrial revolutions. The third transition to a sustainable and emancipated world is every bit as fundamental as those, but will have to be made in a much shorter timescale.

Lest this seem an insuperable task, it is worth putting it in perspective. There are three positive factors to be considered. One is the existence of convincing arguments that, in terms of the personal experience of violence, the world is actually less violent than in recent centuries[17]; more recent experience suggests that the second decade of this century is so far seeing fewer conflicts than the 1990s.

The second is that there are many examples of specific events serving as warnings of greater problems and eliciting the appropriate responses. Two examples from the British experience are interesting here. The 'Great Stink' of London in the hot summer of 1858 made it almost impossible to live or work close to the River Thames, which had become little more than a giant sewer. Parliamentary and other elites were directly affected and finally agreed to long-planned public health works which also greatly reduced the incidence of cholera and other enteric diseases. Nearly a century later around 4,000 elderly, bronchitic and asthmatic people were killed by the four-day 'Great Smog' of London, prompting radical improvements in air pollution control that were already being called for. At the international level, atmospheric chemists and others were already pointing to a potential danger from CFC disruption of the ozone layer, helping to ensure that when the ozone 'hole' was discovered over Antarctica, the intergovernmental response was rapid. In all of these cases there had previously been a process of suggesting the possible and even of planning for the necessary.

Third, if prophecy is indeed suggesting the possible, then there are many good examples of thinking, planning and acting that are already under way relating to the challenges outlined here. To take Britain as an example again, they include the sustained pioneering of the Centre for Alternative Technology's work on renewables and energy conservation and the 'Great Transition' project of the New

[17]Steven Pinker, *The better angels of our nature* (New York: Viking, 2011).

Economics Foundation and the work of the Finance Innovation Lab, as well as the work on sustainable security promoted by the Oxford Research Group.[18] More generally, there is a wealth of activity aimed at promoting more equitable and sustainable technologies and economies right across the world, not least in the global South, and the speed of take-up of relevant new technologies is often remarkable. Also, there are already many examples of economic organization that eschew over-reliance on the profit motive. Across the world there are some 950 million members of cooperatives and other mutual organizations, and in many parts of the world it is feasible for individuals to undertake the majority of their economic interchanges through such organizations.

In conclusion, if the 1945–2045 period really is, as argued here, a particularly important phase in human development, then the last third of that period is the most challenging of all. Thermonuclear war has been avoided, so far, but the weapons still exist in their hundreds and even thousands, and new biological and physical technologies are under development with great destructive potential yet are barely under control. Beyond that is the less obvious but no less fundamental issue of living within biosphere capacity. If progress is made towards a markedly less weaponized, more sustainable and more equitable world during the course of this century, then those alive towards its end may well reach the conclusion that the necessary transformations in behaviour and attitude came to the fore not just in the last third of the period discussed here, 1945–2045, but perhaps even more specifically in the late 2010s and early 2020s. Let us hope they will be chronicled in this journal as it approaches its next milestone.

[18]Centre for Alternative Technology, http://www.cat.org.uk/; New Economics Foundation, http://www.neweconomics.org/; Finance Innovation Lab, http://www.thefinancelab.org; Oxford Research Group, http://www.oxfordresearchgroup.org.uk, all accessed 30 Nov. 2013.

Bradford, Its University and Peace Studies

The Origins of the University

Bradford is an industrial city of a half a million people in the County of Yorkshire in northern England. Its climate, topography and proximity to resources enabled it to come to the fore in the nineteenth century as one of the world centres of the woollen textile trade resulting both in rapid urban expansion and the attraction of immigrants from continental Europe. That feature of inward migration is a key aspect of the city with early groups from Germany, Poland, Ireland and other European countries joined more recently by large numbers from South Asia, especially Pakistan. In the mid twentieth century the wool textile industry went into decline in the face of competition and more rapid modernisation elsewhere in Europe and the wider world and, as a result, the modern city of Bradford has a more diverse economy, including some engineering, chemicals and service industries.

As with other industrial towns and cities, post-school education required specialist training and in 1832 Bradford established a Mechanics Institute from which the modern University can be traced. The Institute evolved into a larger Technical College, established in 1887 and that developed over the following years into a substantial technical institute specialising in science and engineering degrees. In 1957 the Bradford Institute of Technology was formed as one of a number of Colleges of Advanced Technology, mostly based in industrial cities.

In post-war Britain only a very small proportion of the population went into higher education, either in elite universities such as Cambridge, Oxford and St Andrews or a small number of civic universities. There was some expansion in the late 1950s and then, in the mid-1960s, the pioneering Labour government of Harold Wilson embarked on a very substantial expansion in higher education with four elements. One was the opening of entirely new universities, usually on rural sites, with these including York, Lancaster, Kent, Essex and Warwick and the second was new universities based on what had been urban technical institutions but relocated to new campuses, again on rural sites, such as Bath and Surrey. The third group was

P. Rogers, *Paul Rogers: A Pioneer in Critical Security Analysis and Public Engagement*, Pioneers in Arts, Humanities, Science, Engineering, Practice 21,
https://doi.org/10.1007/978-3-319-95150-8

some thirty city-based technical institutes that were developed into polytechnics. They were intended to remain primarily vocational and did not initially have the power to award their own degrees, whereas a few of the larger institutions were designated universities, including Bradford, Salford and Aston. Their emphasis was also on vocational degrees but with more scope for diversification, and their other distinguishing feature was that they retained their urban locations and close connections with their communities.

Bradford University was thus established in 1966 under the leadership of Professor Ted Edwards, a noted educational reformer often known colloquially as "Red Ted". By 2017 the University had grown to over 11,000 students in Bradford and several thousand studying for Bradford degrees across the world. In the years that followed its foundation, the university developed strengths in the life sciences including pharmacy and cancer research, with the School of Health Studies particularly well-known for its pioneering work on dementia. The University is also part of one of the largest research studies in the world, the *Born in Bradford* project which is tracking the lives of 13,500 children born at the Bradford Royal Infirmary between March 2007 and December 2010, and their parents, over 30,000 people in all. Other notable areas of research are materials science and archaeological sciences.

From the early years of the new university, Professor Edwards and his deputy, Professor Robert McKinlay, were intent on using the opportunity to develop new interdisciplinary departments, especially in the social sciences and not least to balance the very strong emphasis on science and technology. These included areas such as European Studies, International Development and one of the first Schools of Management in the UK, and it was in this context that a School of Peace Studies was established, taking in its first students in 1974.

Peace Studies at Bradford

Unlike the other interdisciplinary initiatives at Bradford, the idea of having a Peace Studies department came not from within the university but from an external impetus, members of the Society of Friends or Quakers. In Europe in the early post-war years there had been several initiatives in peace research and peace studies, including, in the 1960s, the Polemological Institute at Groningen in the

Netherlands, the Stockholm International Peace Research Institute (SIPRI) and the Peace Research Institute of Oslo (PRIO). There was no direct substantial equivalent in the UK and a small group of Quakers in the North of England established a group to rectify this.

In the very early 1970s a number of universities were approached, Bradford was particularly interested and the Quakers set up the Quaker Peace Studies Trust to raise funds to establish a School of Peace Studies. This was remarkably successful and the University established the new department in 1973, the first postgraduates enrolling in 1974 followed by undergraduates the following year. Over the past 45 years Peace Studies at Bradford has graduated some 2,000 students many of whom have gone on to work in some of the most troubled places in the world.

In 2017 Peace Studies was united with the development studies centre to create the Division of Peace Studies and International Development and now offers an expanded range of degree programmes. At the undergraduate level these comprise *Applied Peace and Conflict Studies, Global Politics and Development, International Relations* and *Politics and Security Studies.* At the postgraduate level, courses include *Advanced Practice in Peacebuilding and Conflict Resolution, Peace Conflict and Development, Sustainable Development and Peace* and *Resilience and Social Justice*, and there is also a substantial doctoral programme.

Peace Studies welcomes students from all over the world, with many of the postgraduate students coming with years of experience directly related to the work at Bradford. In 2000 Bradford became one of the first of the new Rotary Foundation International Peace Centres, along with universities in Thailand, Japan, Australia, Sweden and the United States.

Over the years Peace Studies at Bradford has been noted for its research on arms control and disarmament, the arms trade, the changing causes of conflict, conflict resolution, human rights and community participation, social movements and social change and areas in conflict. It has long had a reputation for rigorous independent analysis which does not always endear it to government.

Unusually for a university department, Peace Studies at Bradford has its own mission statement which explains how it is more than a standard academic centre:

> We combine empirical, theoretical and applied research with sustained engagement at international, regional, national and local levels to analyse, prevent and resolve conflicts and develop peaceful societies. We aim for an enabling environment for international research excellence involving diverse and critical approaches.

About the Author

Paul Rogers (UK) is Emeritus Professor of Peace Studies at the University of Bradford and Global Security Consultant with Oxford Research Group (ORG). He has worked in the field of international security, arms control and political violence for 40 years, lectures at universities and defence colleges in several countries and is international security advisor to *Open Democracy* and the Oxford Research Group. Paul holds a B.Sc. and Ph.D. from the University of London and an ARCS and DIC from Imperial College London. He is a member of the Royal Society of Biology and has lectured at Imperial College and worked as a Senior Scientific Office in Uganda and Kenya.

Paul has written or edited 29 books and has published over 150 papers with his work translated into many languages. Among his major books are: Hans Günter Brauch, P.H. Liotta, Antonio Marquina, Paul F. Rogers and Mohammad El-Sayed Selim (editors) *Security and Environment in the Mediterranean: Conceptualising Security and Environmental Conflicts,* Springer-Verlag, 2003; *A War Too Far: Iraq, Iran and the New American Century*, Pluto Press, 2006; *Losing Control: Global Security in the Twenty-First Century*, Pluto Press, London, (3rd edition) 2010, and *Irregular War: the New Threat from the Margins*, I B Tauris 2017.

Address: Professor Paul Rogers, Division of Peace Studies and International Development, Bradford University, Bradford, West Yorkshire, BD7 1DP, UK
Email: p.f.rogers@bradford.ac.uk
Website: https://www.bradford.ac.uk/social-sciences/staff-profiles/peace-studies/rogers-paul.php and http://www.afes-press-books.de/html/PAHSEP_Rogers.htm.

P. Rogers, *Paul Rogers: A Pioneer in Critical Security Analysis and Public Engagement*, Pioneers in Arts, Humanities, Science, Engineering, Practice 21,
https://doi.org/10.1007/978-3-319-95150-8

The manufacturer's authorised representative in the EU is Springer Nature Customer Service Centre GmbH, Europaplatz 3, 69115 Heidelberg, Germany. If you have any concerns regarding our products, please contact ProductSafety@springernature.com

Printed and bound by CPI Group (UK) Ltd, Croydon, CR0 4YY
05/10/2026
02228141-0006